THE PLIGHT OF DOLPHINS

When Humans Get In The Way

CURTIS GLENN JACOBSON

ISBN: 1499241089
ISBN-13: 978-1499241082

DEDICATION

Fortunately, the scene on the front cover is one of humans returning a live, stranded dolphin to the Gulf of Mexico, not stealing it from its family and home. This book is dedicated to people that respect and value the sanctity of wild animals and their habitats..

CONTENTS

ACKNOWLEDGMENTS

I want to thank our Heavenly Father, YAHAVEH.

I am very grateful to the staff at the Smithville Public Library in Smithville, Texas.

I would like to thank the following people at Austin Community College: Rick Meyers, Larry Johnson, Judy Sanders, Yvonne Estes, and Steve Ziser. Mike Lyday at the City of Austin. Also, the following at Florida A&M University (Space Life Sciences Training Program): Linda Chamberlin, Anne Moore, and Corinne (Corey) Johnson. I went to each of these persons as a rough stone and in their own way each one knocked off some rough edges and collectively they polished me to an academic gem.

I also want to thank the wonderful staff at the office of Protected Resources of the National Marine Fisheries Service. The people in this government office have always been courteous, patient, and very helpful to me, especially Jennifer Skidmore, Amy Sloan, and Ann Turbush.

1 FIRST ENCOUNTER

I became aware of *the plight of dolphins* in 1989 when a fellow student showed me some literature she received in her biology class about ocean pollution and the tuna/dolphin problem. I could see she was disturbed by the subject and wanted to do something about it. I was emotionally stirred by the pictures I saw and the words I read. I had no idea that my future was about to change direction.

That was during my first semester in college, and I had not yet taken any science classes. In fact, I had dropped out of high school in the tenth grade, so I had no scientific background or interests. However, all of that quickly changed.

I knew dolphins were mammals, but I didn't know the definition of a mammal. I knew they lived in the oceans, but I didn't know how they breathed or what they ate. I was a dolphin illiterate.

Up until then the only exposure I had to dolphins was the old television show, "Flipper" and a single touring dolphin that came to a shopping center two blocks from where I grew up. I was eleven or twelve and remember that dolphin very well.

The shopping center was a great place to ride a bicycle. One summer day before the dolphin arrived, a couple of guys came in a truck to set up the tank. I watched them all day, each day. The day finally came for the dolphin to arrive, and I was right there watching every move of the handlers.

It is weird because I remember watching the dolphin in the tank, swimming circle after circle around the perimeter, but I don't remember being educated about the dolphin. However, knowing what I know now, I feel sorry for that poor dolphin. The tank was no more than twenty feet in diameter and maybe six to eight feet deep. The top half of the tank was made of Plexiglas so you could watch the dolphin underwater.

The trainer made the dolphin do three or four shows a day in the hot sun. He made it jump, flip, whirl, and retrieve things he threw in the water. That dolphin was desensitized to anything outside the circumference of its little tank. Although from time to time, as it circled the perimeter, I could tell it was looking at me. I could see its eye peering through the thick glass as it made its rounds. What it probably was looking for was that little fish that the trainer kept giving it as payment for doing the tricks. I do remember thinking, "What if that dolphin didn't want to participate with that guy all the time?"

That was before dolphins were bred in places like Sea World for complete exploitation. I don't know when, but I do know that it was captured in the ocean and stolen from its natural family. That is just a given fact. Whenever it was captured, it lived its life from then on in one of two places: a small 4x8 tank in the back of a truck, or the tank I saw it perform in. Either way I would say it had a bleak, abused life. Well, that poor dolphin probably died a long time ago from abuse! I'm sure those who exploited that dolphin would say it was one of the first "ambassadors" of its kind! WOE to them.

It cannot be said that without that dolphin I would never have been enticed to pursue the study of the species. That would be incorrect. The fact that I did pursue the study of dolphins says only one thing—God intended it to be so. Furthermore, I do not believe God ever intended for such a wonderful creature of His creation to be confined to live its life in such enclosed quarters and anyone who does that will pay the price.

There is no doubt that dolphins have drawn the interest of humans throughout the recorded history of "man." Although illusive in nature, dolphins have interacted with and intrigued humans time and time again. Humans have repaid them by capturing them, making circus clowns out of them, slaughtering them, even eating them. That kind of human interaction has prompted legislation in the United States and around the world, i.e. the Marine Mammal Protection Act of 1972, etc..

Although I am a student of these beautiful, intelligent creations of God, I shy away from labels like "scientist." Not that there's anything wrong with being a "scientist," but usually I disagree with some of their general statements. Studying dolphins, or any animal for that matter, for scientific information and studying them because you love them are two completely different things. Don't get me wrong, I do not mean there are no "scientists" that love dolphins. I just mean the two different kinds of study produces two different kinds of results: one that speaks only to other "scientists," and one that speaks to the rest of us normal people. Just because a person has a prefix of "Dr." does not mean they are a know-it-all. They may know a lot about one or a few things, but they may also be very confused about everything else.

Nonetheless, research has produced a wealth of knowledge pertaining to dolphins, but I will venture to say that any information other than anatomy and physiology is nothing other than mere speculation; or at most, observations that anyone could make if they would only take the time and effort needed regardless of whether or not they had the prefix of "Dr." in front of their name.

It is my intention to lay out that information in plain English for the simple reason that scientists write for other scientists not intending for the average human to understand what they are saying. It is my hope to stimulate the reader to continue learning about dolphins. I will even provide addresses and websites in the appendices for you to obtain your own information. A better educated person is one who reads between the lines and searches for their own answers.

I will tell you right now that one scientific word I will fight throughout this work and beyond is "anthropomorphism." Anthropomorphism is placing a trait or attribute assigned to humans on an animal. The scientific community has precluded that animals only have the capacity to react to a given situation, not the capacity to conclude or deduce and are totally incapable of feeling emotions. For instance, if I said that my twelve and a half year old dog was sad after my Mom and Dad died just thirteen days apart, some "scientist" would jump and say, "You're being anthropomorphic. Dogs can't feel sadness!" The sad thing about it is, that "scientist" did not know my dog, nor did he/she observe her actions and reactions at any part of her life. The cold, hard fact is my dog was sad or emotionally disturbed.

Furthermore, if I say, "The dolphin stopped, thought about the situation, then turned in a different direction," most scientists will charge me with anthropomorphism because to them animals do not have the ability to think rationally. While I cannot yet prove empirically a dolphin does think rationally or feel emotions, at least I declare it is possible. No scientist has proved to me that a dolphin does not think or feel. I will show you evidence that they do.

As you continue reading, you will undoubtedly sense my biased bend towards conservation rather than exploitation and creation rather than evolution (macro-evolution, that is). I am not ashamed of the fact that I am a Christian, and I write from that perspective.

Ignorance breeds reckless indulgence. Baba Diome said that we save only what we love, we love only what we know, we know only what we learn. May the following pages be a learning experience, the acquisition of knowledge. Come on, let's dive into the secret world of dolphins and see what we can find.

PART ONE: GETTING TO KNOW THEM

Even the most subtle mention of dolphins raises curiosity and intrigue in people. Perhaps it is their illusive nature, which makes them so difficult to observe, or their mythological history that gives us a shadowed perception of where they came from and what they really are. Maybe it is the natural smile God bestowed on their face that captures our attention. Whatever it is, we cannot escape the truth that there are more people endeared to these creatures of the sea than there are those who would rather be alienated from them.

Recorded history provides us with an insight to their behavior and interactions with humans. While I have never seen a report that a dolphin has killed a human, there are stories where dolphins have injured people. After all, they are wild animals. I think it would be safe to say that if you were in the water with a dolphin and it nipped at you that would be its way of saying something like, "I don't like that. Don't do that again!" Or if you're holding its dorsal fin and it begins to dive it is probably trying to let you know it doesn't want to play that game. Your guess is as good as mine.

Without using human languages, animals do communicate with people when the need arises. It is the responsibility of humans to detect and understand that communication. Although God gave us dominion over the earth and animals in the form of stewardship, we become an intruder when we venture into the habitat of wild animals without regard to their natural existence.

2 OUR FRIENDS IN THE SEA

In ancient Greece, dolphins were thought to be gods and to kill one was considered murder. Dolphins were displayed on their coins. Perhaps the oldest known picture of a dolphin is on the island of Crete in the Mediterranean Sea and may be as old as four thousand years. Aristotle studied dolphins and wrote that they care for man and enjoy his music.

You may have heard the old story of the little boy who wandered out into the sea in his skiff. As he rowed further and further the shore grew smaller and smaller. The intensity of the wind increased and changed the course of the skiff. Before long, the little boy was lost and unable to fight the strong current, so he drifted hoping to be rescued. As the day grew longer, a storm developed on the horizon and overtook the little skiff. The little boy was tossed from side to side until the skiff capsized and he was spit out into the cold sea itself. After struggling with the sea, he was overcome by its strength and as he began to slowly sink into the darkness he prayed for help.

As he peered into the depths below, a large, sleek figure appeared and came closer and closer. The strong, graceful figure lifted the little boy to the surface and the two swam away from danger. Back in the safety of his home, the little boy told his family the story of how a dolphin came to his rescue after praying for help and returned him to his familiar shores.

Dolphins are gregarious animals, with their own kind, and rarely travel by themselves for more than just short periods of time. In their natural habitat, dolphins depend on their "pod" for survival. Dr Ken Norris says that in the open ocean, dolphins live a completely free life with no obstacles. "They are dominated by the need to live in societies and a lone dolphin is actually less than a complete animal" (Norris 1992).

Cetaceans (including dolphins) are the only marine mammals that live their entire lives in the water. Other marine mammals, like seals and sea lions, haul-out of the water and spend time on the shorelines.

Let's take a moment and trade places with a dolphin. We're in the water now and there is a plethora of sensations—some are familiar, some foreign. Although our body would weigh maybe four hundred pounds out of the water, in the water it is buoyant and we feel a sense of weightlessness. Our eye sight is very similar to what we are accustomed to as a human, so we look around to get our bearing. In the process, we find the water is no where as clear as the air and our sight is limited. We discover this is a three dimensional world because there are other creatures swimming around below, above, in front, behind, and to our sides. Our brain is even larger than before and it is processing the incoming information accordingly.

As we look above, we see a strange sight; it is smooth and shiny like a sheet of glass and we realize that is where we must go for our next breath of air. Our muscles are no longer producing lactic acid as they burn oxygen so we do not feel that painful desire to take a breath, but our brain is very efficient and it is telling us to make that move. As we move our tail flukes up and down, we discover the power we have to propel us through the water and our pectoral fins act as rudders to steer us in the direction we want to go. As we "fly" through the water, we feel our skin rippling, reducing friction. Finally, the top of our head breaks the surface of the water and we take a breath of air.

Back to our human form. The journey we just took was very brief and we did it without even using what would have been our most important tool—sonar. As a human, most of us probably consider "our" dolphin habitat as a hostile environment because it is completely opposite of what we are accustomed to living in our world the air.

Since dolphins breathe air, they can live out of the water as well. However, out of the water there are several elements detrimental to their survival: gravity and temperature. Without proper support the weight of their body will eventually take its toll on their internal organs. Without the use of water to dissipate their body heat, their skin will begin to dry and crack and they will eventually succumb to hyperthermia.

With all that said, it shouldn't be difficult to get an understanding of how hard it is to study or even observe dolphins in their natural habitat. Standing on a shoreline, you may see heads and dorsal fins breaking the water surface. Or you may be lucky enough to see them breach (jump out of the water), but those two events happen within a matter of seconds out of no where.

Obviously, the best way to encounter a dolphin is to enter the water in their presence. In no way do I recommend doing that. First, because it is illegal to interact, or harass, dolphins in U.S. waters. Second, the instant you hit the water, there probably will not be a dolphin left anywhere within your sight. But if you think you can just go out in the ocean and jump in the water with some dolphins, you better think again. Wild dolphins, in general, do not appreciate the company of humans. Thank God! There are, however, numerous accounts of lone dolphins or groups of dolphins that "seemingly" seek out human contact.

Of course, humans do not need a reason to stick their nose into anything, but what benefit could a dolphin get from an encounter with a human? Why would a wild dolphin swim to a human? They don't eat humans. They can't talk to us. What could it be that a dolphin would want from a human? Let's look at several documented dolphin/human interactions.

Perhaps, the most famous location to encounter wild dolphins is at the Australian beach of Monkey Mia. For many years, wild bottlenose dolphins (*Tursiops truncatus*) have visited the same beach. There are now rangers to mediate the contact between dolphin and human, and I'm sure the dolphins have become familiar with certain people, but for the most part the dolphins are greeted by newcomers each time they visit the beach.

On their own, the dolphins swim up to the beach in water as shallow as a foot deep with their bellies on the sandy bottom and their dorsal side half way out of the water. With their eyes under water, they would see dozens of human shins, ankles, and feet like we would see tree trunks in a forest.

The rangers usually start the visitation equipped with a single bucket of dead fish, allowing a few lucky humans to feed the dolphins one at a time. The amount of fish is certainly not enough to fill the dolphins. During the feeding, the rangers present an interpretation of the history of the event along with some basic information about dolphins and rules for what happens next. Then slowly and gently the humans step out further into the water with hopes that a wild dolphin will come over to them. When the dolphins are ready, they simply turn and swim away to do whatever dolphins do.

Near the islands of the Bahamas is a group of spotted dolphins (*Stenella frontalis*), which doesn't actually seek human contact, but will tolerate humans and even interact with us. According to Denise Herzing, the "dolphins use the shallow (6-12 meter) sand banks in the Bahamas for resting, mating, socializing and feeding" (1990).

Film makers Hardy Jones and Michael Wiese teamed up to produce a wonderful documentary, "A Dolphin Adventure." The team set out to gather new information about human relationships with dolphins. They sought after and found some philosophical answers from global thinker Buckminster Fuller, who theorized that dolphin brains hold information critical to human survival and that they are capable of understanding relationships.

I find this particularly intriguing. First, I must say that it is the mind of God alone that contains information critical to human survival; however, the Bible uses animals in several places as examples for humans to learn from. The book of Proverbs says, "Go to the ant, O sluggard, Observe her ways and be wise, Which, having no chief, Officer or ruler, Prepares her food in the summer, And gathers her provision in the harvest" (NASB 6:6-8). In the book of Matthew, Jesus tells us to "Look at the birds of the air, that they do not sow, neither do they reap, nor gather into barns, and yet your heavenly Father feeds them. Are you not worth much more than they?" (NASB 6:26).

While I do not believe dolphins possess some hidden ability to sway human attitudes, I do believe we could benefit from their behaviors in social relationships. More on that later.

Since dolphins receive about eighty percent of their information from sound, Hardy and Michael engaged the services of Steve Gagne for help in attracting dolphins. Steve designed an underwater piano that worked off the scuba tanks and would reach the lower end of the dolphins hearing range, which is about three hundred thousand cycles per second. After assembling a crew of friends and other dolphin enthusiasts, including whale photographer Jim Hudnall, they set out to find the dolphins.

The first evening, they were met by a small group of spotted dolphins and they all swam together until sundown. The next morning, Steve entered the water with his piano and started playing. Before long, the water was filled with wild spotted dolphins swimming and playing with humans.

There are numerous accounts of lone dolphins seeking or accepting human encounters. Wade and Jan Doak are students of dolphin behavior when it comes to dolphin/human relationships. Wade has several wonderful books out on the subject. He and Jan also founded an organization called Project Interlock. In one of their newsletters Wade writes "I have over forty episodes on file where lone dolphins have sought humans for social stimulus. As I write, eleven are in progress—more than at any time" (Doak 1992).

In the 1950s in New Zealand, a wild dolphin named Opo frequented the near shore waters to play games with children at the beach. Opo even had a favorite little girl to play with, a 13- year-old named Jill.

In the United States, a dolphin named Dolly became friends with a family in southern Florida. Unlike Opo, Dolly had been a "guest" of the United States Navy (released back into the wild) and could understand several commands and play several different games, one of which was to retrieve dimes, no other coins, just dimes.

Dolphins even help fishermen catch fish. In a place called Mauritania, when a school of mullet are sighted children beat the water with sticks, and dolphins drive the mullets towards the beach where the men wait with nets.

Strange as it may be, dolphins are not always welcomed by the public. In his book "Follow The Wild Dolphin", Horrace Dobbs reveals the story of Donald. The bottlenose dolphin (*Tursiops truncatus*) showed up in the waters near the Isle Of Mann in 1972. Donald became known for his playful antics with mooring ties and equipment hanging from unexpectant divers. Most people of the surrounding seaside towns enjoyed Donald's visits, but some fishermen had different ideas. They blamed Donald for damaged nets and low fish counts. Donald was shot in the head, but fortunately he survived and became the center of controversy between the fishermen and animal rights activists.

These kinds of stories, whether good or bad, have the affect of raising public awareness of dolphins. We can only speculate why some dolphins are "attracted" to humans while others maintain their distance. On the other hand, humans are attracted to the free, unencumbered lifestyle of wild dolphins. Dolphins arouse a sense of brotherhood and other feelings of happiness in humans. Of course, that is a general statement and some exceptions will be discussed in later chapters.

3 HISTORY

Dolphins are air-breathing mammals that live their entire lives in water and spend much of their time near the interface between the water and the air. Current scientific "theory" claims that their modern form evolved from land dwelling creatures about sixty-five million years ago. I reject that theory. I believe God created dolphins as a dolphin in the first and second Earth age. I do not believe in macro-evolution!

It is thought by "evolution scientists" those four-legged creatures began searching for food in the shallow sea waters of what we now call the Mediterranean Sea. Finding an abundance of food, they increased their frequency of visits and improved their "fishing" methods over time. As they spent more and more time in the water, their clawed feet started turning into webbed paddles (a favorable change). Since the fishing was good, the creatures nostrils started migrating from the end of their snouts to the top of their heads so they could keep their eyes under the water to increase the amount of time spent searching for food (another favorable change). Over time, the hind legs disappeared and a tail fluke appeared. May I be so polite as to say that is nothing more than BS!

That all sounds wonderful if you are gullible enough to believe the lies that have no evidence for proof. The scientific community has even gone as far as to draw pictures of what these creatures would have looked like before, during, and after the changes they propose took place. That is very thoughtful of them because they have never found one of those creatures to take a photograph of it. They make up in their mind what they "think" that animal would have looked like "if" it had existed, but it did not. Those Ph.D. scientists think you will believe their lies better if you see a picture of an animal that doesn't exist. They are right for the most part because we humans are a gullible lot.

The fossil record continues to remain incomplete. "Incomplete," in that it fails, as of yet, to produce any specimens linking one fossil to another fossil, one in the middle of two, which demonstrate the dramatic "favorable" changes evolutionists need to prove their "theory" of evolution. Therefore, scientists say, "look at this one and look at that one . . . they are different, so there must be some animal in the middle that has characteristics of both all wrapped up in one animal." So they show you one bone, draw an imaginary animal around that one bone, and say to you, "Do you see? I told you there was an intermediate animal." But it is nothing more than a vivid imagination and a lie! They do not want you to believe in God!

This process of "favorable" changes is referred to as macro-evolution and relies on the fossil record for its evidence. Darwin himself stated that his "theory" depended on future scientists finding the fossils of the creatures that would support the "theory." Without conclusive, viable evidence to back up the claims of the "theory of evolution," one must have great faith that Darwin knew what he was talking about because there is no proof. It is a process that replaces specific characters and traits with others that work better and are more efficient, a process where a creature is able to change from what it is or was into a completely different creature.

As a teacher or a professor proclaims the validity of this process, they are quick to install a disclaimer referred to as "acquired characteristics." They say that a giraffe did not acquire a long neck simply by stretching it to reach a higher branch on a tree as French naturalist Jean Baptiste Lamarck suggested in 1809.

Rather, the mechanism in evolution that brings about the long neck of a giraffe is "inherited" (Darwin 1962) variations in reproduction and goes something like this: somewhere in the world a long time ago a baby giraffe was born with a neck that was longer than any of the other thousands or millions of baby giraffes. That baby reached adulthood and could obtain more and better food because of its longer neck. That baby's "long-neck" traits were passed on to one or more of its babies. Over time, there were a lot of giraffes with long necks, and the giraffes with short necks were unable to successfully compete for food and died off. Now all we have are giraffes with long necks. Interesting, isn't it? But not true.

The obvious question is: if the giraffes wanted to get higher up into the trees why didn't they develop wings? After all, scientists claim it was the reptile family that developed wings—reptiles that didn't even eat leaves on trees. All they wanted to do was move around faster and further, so they turned into birds. Well, if transforming from a reptile to a bird was so beneficial why did not all animals turn into birds? Furthermore, why do we still have snakes and lizards crawling around on the ground? Why didn't they turn into birds?

After his voyage on the H.M.S. Beagle (1831 - 1836), Darwin began piecing together his theory of evolution by way of a process he called "natural selection." The observations he made while on that journey around the world produced this "theory" for Darwin:

- **Homology**: structures have an anatomical history which are used for different purposes (arms, forelegs, wings, etc...).

- **Biogeography**: similar habitats in very different parts of the Earth have organisms in similar niches and they resemble each other. Australian marsupials resemble placental mammals: marsupials are not placental, mammals are [an exchange system for food energy between parent and offspring].

- **Embryology**: *Ontology* (development of the individual) *Recapitulates* (goes back over) *Phylogeny* (development of phyla or species). In other words, as a human develops it goes through the same stages as other animals in lower phyla (our ears develop from gill slits and we have a tail that disappears before we are born). Show me a human fetus with a tail.

- **Vestigial Organs**: remnant organs that remain in an organism, which have no meaningful function; (our appendix, a boa's hind legs, whales have a pelvis).

From these observations Darwin concluded:

- There is variation between members within species.
- Some of this variation is inheritable.
- More offspring are born than can possibly survive.
- The survivors are those best adapted to live in their local environment and they will pass on their traits to a greater percentage of the next generation [survival of the fittest].

While "macro-evolution" is an attempt to explain how one creature "turns" into another, micro-evolution explains the differences we see within one species over time. For example, two thousand years ago the average human male height may have been six feet two inches, one thousand years ago five feet eight inches, today six feet even. While the average heights changed over time, we remained humans and did not change into some other kind of animal.

However, if we are to believe that "evolution" is the force from which we were created, instead of by God, then we must face the fact that eventually we will no longer be humans, but some other creature because according to evolution there is no higher purpose for our existence—we simply happened to evolve here by chance. We came from nothing and will end as nothing.

When Charles Darwin invented his "theory of evolution" the field of micro-biology had not yet been discovered. He based his theory on the observable differences he saw in animals around the world then speculated on how those differences came about. While morphological differences are observable with the naked eye, cellular differences are not.

Microscopes of Darwin's day were weak and crude. For this reason, he called the cell simple. Today, we know this is not the case.

The development of transmission and scanning electron microscopes opened a new realm for biologists. Now biologists can not only find and observe individual cells, but can actually study the inner parts contained within. From this newly found power we can begin to understand how inadequate Darwin's theory really is.

In his book, Darwin's Black Box, Michael J. Behe reveals the enormous amount of information, which Darwin had no access to. Behe explains that a single cell has trillions of "machines" and each one of those has another trillion machines in it. Each system of machines is irreducibly complex, meaning it can only work with all of its parts just as they are—take anything away or add something else and the system will fail.

I will leave the technical arguments with the so-called "experts" in each of their own fields of study, but this newly found information makes it rather obvious to me that Darwin was not playing with a full deck. Take the whale for example, based on evolution, scientists say the nostrils migrated from the front of the snout to the top of the head where it is located today. That is one specific characteristic of whales. According to evolution, that one change in body structure was possible only through gene mutations that were favorable (we have gene mutations happening, or trying to happen, in our bodies all the time, but our cells (DNA, RNA) are so complex and well organized that they are corrected before they can cause damage. Most mutations would destroy us before or at birth).

Keep in mind we are talking about only one of the many changes in an animal that because of evolution scientists claim went from being a terrestrial quadruped (four legged) covered with hair, with stand-up ear flaps and a tail to a smooth skinned, legless swimming machine with a blow hole on top of its head and a biological sonar system for navigation and hunting.

Let's examine that one major change in body structure of a Cetacean. Since we have established that mutations are generally detrimental to one's life, how many mutations, or generations of mutations, would have to be successful before the blow hole and all of the nasal passages ended up on top of the head? Would the change have happened all in one generation? What use would be beneficial to an animal if during the "migration" period the nasal opening was positioned between the eyes as it moved from the end of the snout to the top of the head? You can see how silly this sounds, but let's not stop there.

We are still talking about only one major change in body structure and its need. A body may have a successful mutation, which may lead to another and another if it is beneficial, but to have many organ systems changing at the same time to accommodate the initial mutation is preposterous. Did the animal need the hole in the top of its head before it became an efficient swimmer or did it need to replace its legs with flippers before it needed a blow hole?

At what point of the evolutionary process did the animals respiratory system adapt from the needs of a quadruped to that of a swimmer? For example, if you take a deep breath, hold it, and dive into water, before too long your body will begin to hurt and you will feel the need to take another breath of fresh air. That's because as our muscle cells burn oxygen they produce lactic acid, which make the body feel fatigued. But the muscle cells of Cetaceans do not produce lactic acid. Did this complex change also take place at the same time as the migration of the nostrils to the top of the head?

Cetaceans have relatively small lungs. When it takes a breath of air and dives beneath the surface the lungs compress and drive the air into the windpipe and its branches and into the extensive nasal passages. The nasal passages also have thickened membrane linings, which prevent gas exchange to the tissues. Even if the alveoli of the lungs collapse, oxygen can still be exchanged because the terminal airways of the lungs have a unique rich network of capillaries in the epithelium. The increased area provided by the capillaries allows for a rapid return to the lungs of any nitrogen absorbed during a dive. This is thought to be the reason Cetaceans do not experience the bends (Evans 1987). Did all of this take place before or after the nostrils became a blow hole on top of the head because the respiratory system described here is definitely advanced and certainly not found in terrestrial animals?

We now know that the purpose and task of DNA is to replicate itself as a template for further reproduction. DNA tells all the different parts of each and every cell what to do. I was taught in college that I was the product of evolution—"from ooze, to the zoo, to you," so to speak. But the very first cell that formed out of the primordial soup did not even have DNA, or a nucleus for that matter. Those professors wanted me to believe that evolution created DNA so it could run the "show" from that point forward. They wanted me to believe that DNA is an intelligent creative force, but of course that would be anthropomorphic.

However, DNA is so "smart" that if it makes a mistake in replicating itself it detects the mistake, locates it, then corrects it. Failure to correct the mistake leads to the destruction of the strand. Ask yourself, how many birth defects would you consider as "beneficial" for your own newly born baby? Do you want your kid to be born with Downs Syndrome? Do you want it to have six fingers on each hand, or no legs, or a head that is twice the size of its body? Common sense will tell you that mutations and defects are not beneficial to the well being of your baby.

Evolution is based on gradual, beneficial changes over long periods of time. I do not consider the relocation of nostrils from the front of the snout to a single blow hole on top of the head to be a gradual change or beneficial unless all the other changed systems of respiratory and nasal passages were already in place, but why would that have happened if the blow hole was not in place. What we have here is an endless quandary that can never be explained by "evolution!"

It is time to face the facts, folks. The only reason any human being chooses to believe in evolution is because they have chosen to reject divine creation by an almighty God. He created trees as trees, flowers as flowers, birds as birds, and people as people.

We have all heard of the process of metamorphosis in which a fertilized butterfly egg turns into a caterpillar, a chrysalis, and ends up being a beautiful butterfly that looks like its parents. A worm turning into a butterfly sounds like evolution to me. It goes through several gradual changes over a short period of time and comes out looking like something else. It happens to every butterfly and has been happening for a very long time.

I have personally watched, studied, and documented the event more than once and each time the same thing happens—a worm turns into a butterfly, but my college professors were quick to reject the idea that that was evolution. I wonder why? Evolution is a lie! That's why.

The differences between a mammal with two legs, four legs, or two flippers were miraculously created by God! The differences between a mammal that walks on land or one that swims in the ocean were miraculously created by God! The differences between you and me were miraculously created by God!

Darwin's theory of evolution is a farce. There is no scientific evidence to support the idea, which was simply made up in the mind of a misdirected human. To believe in evolution requires a greater amount of faith than it does to believe in divine creation by God. I choose to reject evolution and believe the word of God.

CLASSIFICATION

God created many different Cetaceans ranging in size from the 1.5 meter harbor porpoise (*Phocoena*) to the largest creature to ever live on the Earth, the blue whale (*Balaenoptra musculus*) at 30.5 meters..

While the words dolphin and porpoise have been loosely interchangeable, there is a distinct difference between the two. There are six species in the true porpoise family, Phocoenidae. Generally speaking, porpoises are much smaller than dolphins on average and have shorter less pronounced rostrums (snouts). On the other hand, the widely known killer whale (*Orcinus orca*) is actually a member of the dolphin family, Delphinidae. Hopefully, the following information will help clear up some of the confusion.

A breakdown of the classification of bottlenose dolphins goes like this: KINGDOM Animalia, PHYLUM Chordata, SUBPHYLUM Vertebrata, CLASS Mammalia, SUBCLASS Eutheria, ORDER Cetacea, FAMILY Delphinidae, GENERA *Tursiops*, and SPECIES *truncatus*.

The ORDER Cetacea contains three SUBORDERS: Archaeoceti, Mysticeti and Odontoceti. Archaeoceti is believed to be the ancient, first recognizable "type" of Cetaceans all of which have been extinct for a long, long time. If you choose to believe in evolution, this is the first group of creatures to show the body structure after all the changes brought about by evolution to result in the form of whales. If you believe that, then you must answer this question: did the ancestors of Archaeoceti choose to become water dwelling creatures or were they forced to do so? If you can prove the answer to that question beyond a reasonable doubt, you will be a candidate for the Noble Prize in Biology. Good luck with that!

I reject the theory of evolution and offer a more reasonable and believable explanation. God created the ancient animals such as the dinosaurs and Archaeoceti in the first Earth Age millions of years ago. At that time we were here in spiritual bodies, not in the flesh. Those ancient animals were in flesh bodies. That is why we can now discover their bones. At the Katabole (the overthrow of Satan) God chose to destroy the first Earth Age and those animals became extinct. Then God created this second of three Earth Ages and created the animals we now have with us. In this Earth Age all the animals, and we humans, are as God created them. The Bible declares the Earth is very old. We do not know how old it is.

The remaining two suborders of Cetacea are extant. These are the whales God created in this second Earth Age. Mysticeti are the whales with baleen instead of teeth and have two blow holes. Odontoceti are the whales, dolphins, and porpoises that have teeth and one blow hole.

The following is the taxonomy for the Order Cetacea:

Suborder: Archaeoceti (extinct)

Family: Protocetidae

Group 1: Protocetus atavus, Pappacetus lugardi

Group 2: Eocetus schweinfurthi

Family: Dorudontidae

Group 3: Dorudon, Zygohiza

Group 4: Keneodon, Phocoetus

Family: **Basilosauridae**

Group 5: Prozeuglodon, Basilosaurus

Group 6: Platyophys

Suborder: Mysticeti (The Baleen Whales)

Family: **Aetiocetidae and Cetotheriidae are extinct**

Balaenidae (The Right Whales)

Balaena mysticetus (Bowhead Whale)

Eubalaena glacialis (Northern Right Whale)

australis (Southern Right Whale)

Caperae marginata (Pygmy Right Whale)

Balaenopteridae (The Rorqual Whales)

Balaenoptera musculus (Blue Whale)

physalus (Fin Whale)

borealis (Sei Whale)

edeni (Bryde's Whale)

acutorostrata (Minke Whale)

Megaptera novaeangliae (Humpback Whale)

Eschrichtiidae (The Gray Whale)

Eschrichtius robustus (Gray Whale)

Suborder: Odontoceti (The Toothed Whales)

Family: Agorophiidae and Squalodontidae are extinct

Physeteridae (The Sperm Whales)

Physeter macrocephalus (Sperm Whale)

Kogia breviceps (Pygmy Sperm Whale)

simus (Dwarf Sperm Whale)

Monodontidae

Monodon monoceros (Narwhal)

Delphinapterus leucas (Beluga Whale)

Ziphiidae (The Beaked Whales)

Berardius bairdii (Baird's Beaked Whale)

arnuxii (Arnoux's Beaked Whale)

Hyperoodon ampullatus (Northern Bottlenose Whale)

planifrons (Southern Bottlenose Whale)

Ziphius cavirostris (Cuvier's Beaked Whale)

Tasmacetus shepherdi (Tasman Beaked Whale)

Mesoplodon densirostris (Blainville's Beaked Whale)

bidens (Sowerby's Beaked Whale)

europaeus (Gervais' Beaked Whale)

mirus (True's Beaked Whale)

layardii (Strap-toothed Whale)

grayi (Gray's Beaked Whale)

bowdoini (Andrew's Beaked Whale)

pacificus (Longman's Beaked Whale)

hectori (Hector's Beaked Whale)

ginkgodens (Ginko-toothed Beaked Whale)

stejnegeri (Stejneger's Beaked Whale)

carhubbsi (Hubb's Beaked Whale)

Delphinidae (The Dolphins)

Orcaella brevirostris (Irrawaddy Dolphin)

Peponocephala electra (Melon-headed Whale)

Feresa attenuata (Pygmy Killer Whale)

Pseudorca crassidens (False Killer Whale)

Orcinus orca (Killer Whale)

Globicephala elaena (Long-finned Pilot Whale)

macrohynchus (Short-finned Pilot Whale)

Steno bredanensis (Rough-toothed Dolphin)

Sotalia fluviatilis (Tucuxi)

Sousa chinensis (Indo-Pacific Humpbacked Dolphin)

teuszii (Atlantic Humpbacked Dolphin)

Lagenorhynchus albirostris (White-beaked Dolphin)

acutus (Atlantic White-sided Dolphin)

obscurus (Dusky Dolphin)

obliquidens (Pacific White-sided Dolphin)

cruciger (Hourglass Dolphin)

australis (Peale's Dolphin)

Lagenodelphis hosei (Fraser's Dolphin)

Delphinus delphis (Common Dolphin)

Tursiops truncatus (Bottlenose Dolphin)

Grampus griseus (Risso's Dolphin)

Stenella attenuata (Spotted Dolphin)

plagiodon

frontalis

dubia

coeruleoalba (Striped Dolphin)

longirostris (Long-snouted Spinner Dolphin)

clymene (Short-snouted Spinner Dolphin)

Lissodelphis peronii (Southern Right whale Dolphin)

borealis (Northern Right Whale Dolphin)

Cephalorhynchus heavisidii (Heaviside's Dolphin)

hectori (Hector's Dolphin)

eutropia (Black Dolphin)

commersonii (Commerson's Dolphin)

Phocoenidae (True Porpoises)

Phocoena phocoena (Harbor Porpoise)

spinipinnis (Burmeister's Porpoise)

sinus (Cochito)

dioptrica (Spectacled Porpoise)

Phocoenoides dalli (Dall's Porpoise)

Neophocaena phocaenoides (Finless Porpoise)

Platanistidae (River Dolphins & Franciscana)

Platanista minor (Indus Susu)

gangetica (Ganges Susu)

India geoffrensis (Boutu)

Lipotes vexillifer (Beiji)

Pontoporia blainvillei (Franciscana)

Taxonomy, the science of classifying organisms, is open to interpretation. It is possible that "man" has not yet seen every living or dead creature on the Earth. It is also possible that wild animals can cross-breed or humans can cross-breed animals—remember the mule, which is a cross between a horse and a donkey. Although mules are sterile, they still exist. There may also be so-called "biologists" around the world who like to play God and make themselves famous by trying to create new species.

Successful cross-breeding automatically brings into question whether or not the two donor species were classified correctly in the first place. After all, the definition of a species is the ability of a male and female to produce viable offspring. In other words, if those offspring reach reproductive maturity and successfully produce more viable offspring, then they are of the same species. A mule is not a separate species, it is a hybrid.

The Earth is full of varieties and hybrids. Have you been to a botanical nursery or a zoo lately? If so, you probably noticed many different groups of plants or animals that were similar to one another, but yet had something that was different enough to set them apart from the others. The world does not exist in its current state because we humans tell it what to do. A flower does not know that a human has said of it, "You are this kind, that one is another kind." Some humans like to think that's the way the world works, but I have news for them—it isn't.

Hypothetically, if I stand in a boat among a group of dolphins and say, "That one is short and fat and light colored, and that one is long, skinny, and dark colored, and since I have a Ph.D. I declare that they are different kinds of dolphins," would that make me necessarily correct?

We humans come in a variety of sizes, shapes, and colors. We are short and fat, short and skinny, tall and fat, tall and skinny. We have brown, blond, and red hair. We have white - white, pale - white, olive, light brown, and very dark brown skin. In spite of all the variations, we are still the same species (*Homo Sapiens sapiens*). How do we now know this? Because a Caucasian can mate with a Negro and produce a baby that can eventually mate with an Indian, which can eventually mate with a Mexican, and so on. Regardless of our size, shape, or color we produce viable offspring.

You can see that variations are abundant within our own species. Why would that be any different with any other species? And so, the task of classifying organisms becomes very difficult.

4 LIFE CYCLE

As the baby dolphin slipped into its new life in the ocean it lost all the comforts of the home it was used to for about a year in its mother's womb. Dolphins are born tail first. In a matter of seconds the warmth it enjoyed turned to cold. All of a sudden there were no boundaries limiting its movement, and the baby began flipping its tail trying to learn how to swim for the first time. The mother quickly turned and placing the "flat" of her rostrum on the baby's soft underbelly, gently and steadily raised the baby to the surface for its first breath of air.

With her aunt just a few feet away ready to take action at the hint of trouble, the newborn dolphin quickly finds her orientation and the three slowly swim on in the darkness of the night. From that point forward, the mother will rise to take a breath of air each time the baby does even if she doesn't need to.

Dolphins are viviparous, that is, they begin reproduction with internal fertilization. The embryo goes through a gestation period of about a year, taking its nourishment directly from the mother and is born at an advanced stage in development able to learn quickly. The process usually produces one young at a time, and the mother waits about two years before starting the process again.

The mother raises her young without the help of the father, perhaps with the occasional "baby sitting" by other family members or siblings. After birth a baby dolphin receives fat-rich milk from its mother for fourteen to nineteen months, which occurs about every twenty minutes, twenty-four hours a day. When the baby is ready for milk, it nudges the mother's teats and the milk is squirted into its mouth.

New born females stay with their mothers four to six years before leaving her side, if indeed they do leave her. Some females stay with their mothers and her relatives for life. Males, on the other hand, tend to leave their mother's side on their own after four or five years or face rejection by the pod. This is God's way of preventing inbreeding within the pod.

A baby dolphin has to learn how to live its life from the adults it associates with. It has to learn how to echolocate, search for and capture food, and the "do's and don'ts" of social life.

Females usually reach sexual maturity around the age of eleven, males at about twelve, and the average life-span is about twenty-five years (Evens 1987).

SOCIAL STRUCTURE

Dolphins are social animals. Each individual forms bonds with other family members, members of the pod, and as they grow older members of other pods. The first bond a dolphin forms is with its mother at birth. An infant will stay directly at its mother's side for the first four to six months feeding, watching, and learning.

I will go a step further and say that an infant may be able to easily recognize its mother's close associates after birth, especially other females. Dolphins are very intelligent and with their biological sonar one can tell when another is pregnant and at what stage. Females have been known to gather near one who is about to give birth and a human can speculate many reasons why.

First of all, dolphin sonar and echolocation tends to soothe muscle tension as several pregnant human females have noted after being in the water with dolphins. The women noted that a dolphin would single them out of a crowd, come near to them and direct their beams of sonar at the fetus in their womb. It is believed that the biological sonar gives a mental picture similar to medical ultra-sounds performed by human doctors, only better. One can only deduce that since dolphins do this to humans they must also do it to one another in the wild. From a sonar scan, a female dolphin may be able to gain vital information such as the size of the fetus, proper position in the womb, the health of the fetus, even its sex.

Secondly, females have been observed assisting in the birth of babies by using their rostrum to massage the baby through the birth canal and to shove a baby to the surface when the mother was unable to do so herself. These types of acts demonstrate that dolphins display perhaps the strongest and most important of social skills—compassion and altruism (Altruism is defined as unselfish concern for the welfare of others or selflessness). Perhaps we could learn a lesson from dolphins in that matter.

Since there are no humans smart enough to decipher natural dolphin language, there is no way of telling what kind of information is exchanged from "standby" dolphins to a fetus still in the womb, if any. All I am saying is that it is possible.

Furthermore, concerning a baby that stays in the womb for twelve months, since dolphins live in water where sound travels four times faster than in air and many times farther, there is no doubt in my mind that a dolphin fetus can hear the sounds of its mother as well as the sounds of her associates for quite some time before birth. After birth, it can then recognize the sounds and with sight tie the sound to a specific individual.

As a baby dolphin grows, it begins to interact with others of its own age and those below and above its age. With those of its own age, it will play, learning and exploring its power of locomotion and association. It begins to form bonds with individuals other than its immediate family members. Together they begin to learn the parameters of time and space, what is acceptable and un-acceptable behavior, and how to forage for food.

From older individuals, it learns how to socialize with the opposite sex and how to mate. They learn how to defend against dangers like other predators in the area or direct attacks. These kinds of lessons are learned by trial and error and take several years to master.

Both young and older females may stay in the same pod for their whole life, while males are usually transient between groups forming their own life long bonds interacting for purposes of mating and defense against predators. Even males of different groups will come together to fight a shark or a killer whale if needed. Dolphins are so smart that they know when the danger is too great and instead of direct battle they will herd the group and lead them to safety.

As they continue to grow, a dolphin learns gestures such as jaw clapping and head twitching—signals they use to let others know just how they feel.

Dolphins do have their own language, which varies from pod to pod and geographical regions around the world. As I said before, a dolphin fetus is exposed to the communication used by its pod from the moment its organs develop before birth. The dolphins' world is very harsh and full of danger.

Communication is the key to a successful life for each individual. Each pod of dolphins has its own dialect, which it uses to distinguish itself as a group. Groups that frequently interact and mingle together may, and probably do, have similar sounds and gestures intertwined within each dialect for the purpose of determining friend or foe. The amount of interaction between groups determines similarities in the different dialects. This pertains to both genders. However, males travel much further and more often from their group than do the females.

As in human society, dolphins do not tolerate deviant behavior, and discipline is usually administered by the older members of the pod. Dolphins have bullies just like we do. They also have a pecking order, which comes with age. A large part of the time discipline is related to juvenile sexual behaviors. Dolphins possess innate abilities to protect against inbreeding. Typically, the males learn this at a young age and know when to leave the group, either for its own safety as larger males from other groups come in to interact with the females of his group or to go find a female of his own. If he can't control himself among the females of his own group the older females kick him out.

There are very few animals in the world other than humans who engage in sexual activities for reasons other than procreation, dolphins are one of those. It is thought that dolphins use sexual play as a means of defining relationships.

Dolphins rarely travel alone. They are commonly seen in groups of two and sometimes congregate in groups of fifty or more.

We usually don't hear about, or see, dolphins fighting, but it does happen. While I was studying dolphins in the area of Port Aransas, Texas, I was fortunate enough to witness an outright battle between four to eight dolphins in two groups.

There is a causeway that runs seven miles between Aransas Pass and Port Aransas. On the north side of the causeway is a dredged channel (named Aransas Pass) for shrimp boat access to and from Aransas Pass (the town) and the Gulf of Mexico. Port Aransas is located on Mustang Island. The island is separated from the mainland by a ship-channel which provides a commercial shipping lane from the open Gulf of Mexico to Corpus Christi. This ship channel is the only water exchange access between the bay and estuary systems for thirty-five miles north or south of Port Aransas, which means it is also the only access between the gulf and the bays for fish, dolphins, and other aquatic organisms. There is a free ferry system from the end of the causeway to Port Aransas.

With all that said, on this particular day I crossed on the ferry to the causeway for an afternoon of leisurely fishing in Aransas Pass, but when I got there it was my camera I used instead of my rod and reel.

What I found was a couple of dolphins swimming casually about ten feet from the bank. It took me a few minutes to ready my camera. The dolphins stayed.

As I surveyed the situation, I noticed other dolphins in the area. At this point I still had no idea what was about to happen. I was just happy to have some dolphins hanging around close enough for me to watch. Before long I started seeing an elevation in their activity, which intensified rapidly. I had seen dolphins playing before, but this scene had the appearance of something I had never seen—a battle (see Figures 1-4).

Figure 1

Figure 2

Figure 3

Figure 4

As the dolphins came together, I could see them making quick turns and rushes towards each other. There were several breaches and then the water calmed down. After a short moment, the surface exploded. The whole thing lasted about fifteen to twenty minutes and was very violent. I found myself wishing there was something I could do to stop it, but of course, there wasn't. I don't like to see dolphins fight, but I realized I was fortunate to experience such a thing. It was one of those moments when I was able to recognize something I had read about, happening right in front of me. I thank God for putting me in the right place at the right time.

Dolphins are difficult to study; they live in water and come to the surface only briefly for air. Secondly, if you are fortunate enough to be in the water with them, they can vanish with the flip of the tail never to be seen again.

I am very glad to say there is a new "breed" of scientists who have devoted their lives to studying dolphins in their natural habitat. This type of research requires years and years of time and is very expensive. Nevertheless, these scientists have realized the importance of studying dolphin behavior in the wild as opposed to capturing them and bringing them to a swimming pool laboratory, which costs even more. The more you search for information on dolphins, the more you will find the names of these people donning the pages of news paper articles, periodicals, research papers, and books. Let's take a look at some research conducted by both sides of the issue.

PART TWO: RESEARCH

There is an increasing amount of research on dolphins by scientists and enthusiasts as well. The question of "why" is obvious. Dolphins are mystical and illusive. They live their lives with no apparent regard for humans. Yet when you stand on a shore watching them go about their business you cannot help but wonder what makes them tick.

Since the 1950s, scientists have indulged in almost every kind of conceivable research on dolphins one could imagine. From sonar to intelligence, social behaviors to feeding, anatomy, locomotion, physiology, and history—you name it and there is probably a scientist trying to find an answer to it.

We have enclosed them in tiny swimming pools, cut them open, followed them in boats, and jumped in the water with them trying to figure out what makes them tick.

As I stated before, I am of the belief that humans can learn, to some degree, about living life from dolphins. However, I disagree with some of the practices performed by some so-called scientists in order to find the answers. I disagree with slicing open a live dolphin to see what's inside. I disagree with electric shock to see how they will respond. I disagree with confining them in a concrete pool so the public can see how well they can perform. I disagree with teaching them to manipulate explosives in wartime activities!

Only human arrogance says "They are ours to do with however we see fit." Let us not forget that God created dolphins before He created humans. After He created humans, then He gave us dominion over all the Earth and commanded us to be its stewards.

Certainly I am not against dolphin research, but there is a line that must be drawn between finding out what makes them tick and training them to perform as a human wants them to perform! Furthermore, they should not have to forfeit their God-given freedoms in order for us to study them.

There are already way too many dolphins doomed to a life of captivity. Some of them have never seen the open ocean. Some of them have never seen more than one or two other dolphins in their lives. Is that not cruel? There should never, ever have to be one more dolphin captured and confined to a concrete swimming pool!

5 EARLY RESEARCH

God only knows how long humans have studied Cetaceans scientifically or otherwise. In my research, I have viewed many references and the very early studies seem to have focused on traits and characteristics observable from a ship or on land. Things like identification, classification, adaptations for living in water, and migrations.

Here is a brief, inconclusive progression in Cetacean research. In 1877, D.J. Cunningham (Cunningham, D.J. The spinal nervous system of the porpoise and dolphin. J. Anat. II: 209-28. 1877) wrote about the nervous systems in porpoises and dolphins. In 1901, W.B. Benham (Benham, W.B. On the larynx of certain whales. Proc. Zool. Soc. London I: 287-300. 1901) wrote about the larynx in particular whales. In 1928, R. Kellogg (Kellogg, R. The history of whales. Their adaptations to life in water. Quart. Rev. Biol. 3: 174-208. 1928) took up the subject of adaptations for life in the water - followed by A.B. Howell on the same subject in 1930 (Howell, A.B. Aquatic mammals;

their adaptations to life in the water. Springfield, Ill.: Thomas, 1930).

In the 1940s and '50s, there was an upsurge in research related to both biological sounds emitted in the world's oceans and their detection. This phase of research was primarily initiated by the U.S. Navy (Johnson, M..W. Underwater noise and distribution of snapping shrimp with special reference to the Asiatic and Southwest and Pacific areas. PB 40788 (University of California Division of War Research Rept. U146), pp. 1-7. 1944) as a means to combat the effectiveness of German U-boats in World War II.

As the Navy's "call" for information spread across the country, college professors and scientists came out of the woodwork to capitalize on the generous grant money provided by the military.

During World War II, detecting and identifying oceanic sounds was a growing problem and the only known method at that time was to listen by way of underwater audio equipment. They would listen for engines and propellers and try to estimate distance and size of the vessels, but the problem was that some sounds were un-identifiable. "0810. Sound picked up unusual noise . . . could see nothing through periscope on that bearing. The noise sounded like hammering on steel in a non-rhythmic fashion..." (Kellogg, 1961).

Such sounds are emitted naturally by some crustaceans like shrimp and echolocation clicks of Cetaceans. Therefore, the task of researchers was to develop a new system for navigating the oceans and identifying the sounds therein, and they began by studying shrimps, bats, and dolphins. The result was the development of Sound Navigation And Ranging (SONAR).

Investigations and research projects sprang up all across the country. It seems that the U.S. Government would pay big-bucks to anyone who was able to write a proposal for a government grant. All they had to do was to include in their proposals any word like shrimp, bat, dolphin, sound, noise, navigation, navy defense, or anything the navy thought would help gain advantage over the enemy. I have to say that despite the intentions, most of the early research was legitimate and led to the spin-offs, which have supplied the wealth of knowledge we have today; although, we have still only scratched the surface as far as figuring out what makes a dolphin tick.

A man named John Lilly was one of those early takers of government grants. However, to say his research was "controversial" would be a gross understatement. I do not intend to judge the man's mind or heart, only to say his methods for gaining information was far too aggressive and unscrupulous. Although he may have determined, somewhat, that a dolphin's brain is larger than a human's or the vision field of a dolphin's sight or gained an idea of a dolphin's abilities in echolocation, he did it at the expense of the lives of untold numbers of dolphins.

All you have to do is read one of his books to know he sliced open live dolphins. We just don't know how many dolphins he killed that he didn't write about. Furthermore, he learned the hard way that dolphins are involuntary breathers; that is, their brain must tell their lungs when to take a breath—unlike ourselves who do not have to think about when to breathe. Therefore, in order for a dolphin to be put under anesthesia, he must first be attached to an alternative breathing apparatus or he will die!

You are probably asking, "then how do they sleep if they have to think to breathe?" When a dolphin sleeps, which is only about a total of a couple of hours per day, they only "sleep" one half of the brain at a time while the other half continues to run the body (they must stay afloat, continue to breathe, and stay alert from attack of any kind), and they alternate brain hemispheres for maybe twenty minutes at a time.

Nonetheless, you should be assured that methods involving placing live dolphins under the knife have since been guided by ethical scientists.

I have never agreed with the navy's physical use of Cetaceans for explosives retrieval or placement or as so called "watchmen" for sensitive areas around top secrete installations. Such tactics definitely fall under the category of exploitation. While Cetaceans are surely smart enough to carry out the tasks, they are never shown the results of the consequences, which they may or may not be able to comprehend. For example, a wild dolphin will probably encounter sharks in its lifetime. It will learn from experiences of certain situations that sharks are deadly; thus, it will from that point forward reserve a certain respect for sharks and will be able in the future to discern which actions are necessary for a particular situation involving sharks. A wild dolphin may occasionally play with a shark. It may flee from a shark or it may stand against a shark in defense of itself, its mate, or young. The point is, life demonstrates different situations in which lessons may be learned, but the navy demonstrates only the task at hand—not the consequences.

What about Army and police dogs that help locate explosives? Those animals are accompanied by humans. The dog's noses detect the danger, and their trained reaction informs the humans. Therefore, the human and the animal stay alive. Navy dolphins work alone and do not know the consequences.

I think that if the navy demonstrated to dolphins what an explosive was capable of doing, dolphins are smart enough to refuse to cooperate from that point forward! If that were not true, why are dangerous tasks like retrieving unexploded bombs not assigned to humans. The answer is humans know the danger while Navy dolphins are shielded from that knowledge.

It is easy to teach a dolphin to retrieve a ball, a stick, or a Frisbee, not to mention a hand-grenade. Then, if the trainer pulls the "pin" on the hand-grenade and throws it for the dolphin to retrieve, the dolphin will go get the device and as it returns to the trainer it will be blown to bits!!! Come on, people. You can demonstrate this with your own dog if you have one that retrieves and almost every one will if there are taught (do not be a fool by trying something stupid with your dog to see if what I said is true. Your animal is loyal to you and depends on you for food, love, and companionship. Reason this out in your mind, not by experience!).

Animals build relationships with humans based on trust. Usually, they are dependant on the human for food and confined to a particular perimeter by some sort of physical device like a fence. While it is sometimes stated the dolphins are free to leave if they want and choose to stay, it may be interpreted this way: if you were provided constant, free room and board with special attention would you want to leave?

The following chronology of the Navy's Marine Mammal Program was gathered from the internet (www.pbs.org/wgbh/pages /frontline/shows/whales/etc/navychron.html):

1960's: Navy begins use of marine mammals

1965: The Marine Mammal Program began its first military project: SEA LAB II. Working in the waters of La Jolla, California, a bottlenose dolphin named Tuffy completed the first successful open ocean military exercise. He repeatedly dove two hundred feet to the SEA LAB II installation, carrying mail and tools to navy personnel. He was also trained to guide lost divers to safety.

1965-75: The Navy sent five dolphins to Cam Rahn Bay (Vietnam) to perform underwater surveillance and guard military boats from enemy swimmers. Although during this era rumors circulated about a "swimmer nullification program" through which dolphins were trained to attack and kill enemy swimmers, the Navy denies such a program ever existed.

1975: The Navy began training sea lions to recover military hardware or weaponry fired and dropped in the ocean. The sea lions recovered objects at depths of up to 650 feet. The Navy also began training Beluga whales, which could operate in much colder and deeper water than dolphins or sea lions.

1986: Congress partially repealed the 1972 Marine Mammal Protection Act so the Navy could collect wild dolphins for "national defense purposes." The Navy began its own breeding program.

1986-88: The Navy sent six dolphins to the Persian Gulf, where they patrolled the harbor in Bahrain to protect US flagships from enemy swimmers and mines, and escorted Kuwaiti oil tankers through potentially dangerous waters. One dolphin, Skippy, died from bacterial infection.

The Navy abandoned its protection by dolphins of the Trident Missile Base in Bangor, Washington due to a suit filed by activists under the National Environmental Protection Act, claiming the northern waters were too cold for the dolphins, which were captured in the Gulf Of Mexico.

1990's: With the end of the Cold War, the Navy's budget for the marine mammal program was drastically reduced and they began downsizing.

While the U.S. Navy continues their research on dolphins today, it should be noted that they have initiated a program in which certain captive animals may be rehabilitated and released back into the wild. In October of 1993, the NAVAL COMMAND, CONTROL AND OCEAN SURVEILLANCE CENTER RDT&E DIVISION (see Appendix A) published its Technical Report 1549 (Reintroduction to the wild as an option for Managing Navy Marine Mammals).

"1549" reported the findings of two conferences held by Navy commanders and contemporary dolphin scientists in San Diego. In the "EXECUTIVE SUMMARY" of the opening pages of the report, the objective of the conference is stated as ". . .develop training procedures which will allow mammals which are no longer required for this project to be released back into their natural habitat . . ." On the same page under the sub-heading of "CONCLUSIONS" the report says, "There is no compelling scientific reason for reintroducing non-endangered species."

The reintroduction of captive Navy dolphins is hampered by only one thing—money! There is no doubt the Navy knows how to proceed with the release of excess dolphins. The cost of maintaining the dolphins in captivity is about $2 million per year. They estimated the cost of retraining the dolphins could be as much as two to five times that amount. The U.S. Government is obviously willing to pay the price for capturing, training, and maintaining dolphins, but not willing to pay the price for returning them to their natural home. This also includes various species of whales and pinnipeds.

6 INTELLIGENCE

In his book, FOLLOW THE WILD DOLPHINS, Dr. Horace Dobbs (1982) gives us a wonderful story about a wild bottlenose dolphin named Donald. Dr. Dobbs followed Donald from 1972 to 1978 between Scotland, Ireland, and England. On one occasion, Dobbs wanted to film Donald. The water was rough and without realizing it the camera slipped off his neck and sank to the bottom. While Dobbs was searching for the camera, he came face to face with Donald. Through his mouthpiece, he told Donald the camera was lost and he couldn't take his picture. Donald shook his head several times and swam off.

As Dobbs watched, Donald came to a stop vertically on the sea floor, flexed his body a few times and waited for Dobbs. There, beneath Donald's snout, was Dobbs' camera. Somehow, Donald knew what Dobbs was looking for. Did Donald understand what Dobbs had said through the mouthpiece or did he just put "2-and-2" together after, perhaps, seeing the object sink away from Dobbs and then seeing Dobbs wandering around aimlessly? Does this story begin to demonstrate dolphin intelligence or even abstract thought?

Intelligence is defined as the capacity to acquire and apply knowledge—the faculty of thought and reason; showing sound judgement and rationality. For a dolphin that means being able to figure out how to live in the environment, i.e. finding food, navigating the three-dimensional world of water, detecting trouble and danger, even getting along with other dolphins.

The sum of intelligence is the combination of brain functions. That is to say, it would seem that the more operations required of a brain to perform would increase the physical size of the brain and subsequently its intelligence. That may or may not be the case, but the answer is not that simple.

Throughout the various animal phyla, organisms are endowed with different means of surviving in the environment. Each with their own systems of organs and body plans, whether or not they enjoy the five senses bestowed on us humans. Each one lives in the environment according to its God given abilities.

A sponge, having no specialized nerve cells at all, manages to find a suitable attachment site, forms a colony, and performs its function in life—a stationary filter feeder, it eats and reproduces.

Cnidarians, like jellyfish and sea anemones, have no brains or even clusters of neurons, but receive stimulus anywhere along the body, which sends nerve impulses to other regions.

Even at the simplest levels of organization, these animals are successful in their lifestyles and life cycles gathering information about the environment and using it to their benefit. But are they intelligent? They certainly are intelligent enough to survive in the ever changing, harsh environment of the oceans of the world.

Unlike several of my college biology teachers, I refuse to use the terms “higher” and “lower” when referring to different organisms in a phylogenetic tree. Such terms imply that one is actually better than another when speaking in general terms. It is sufficient to say that one has a “more” or “less” complex body plan than the other, not to attach a degree of importance as to whether one is better than the other in the overall scheme of things. Even the least complex of all organisms has an important role to fulfill in the world. After all, they are as God created them, and He does not make mistakes.

Obviously, the more complex an animal’s body plan the more complex its nervous system will be and ultimately its control center—the brain.

A round worm has sensory cells in the front of its body. Therefore, it moves through the soil in a forward motion. Whereas, a flat worm has concentrations (clumps) of nerve cells (ganglia) throughout the length of its body. Ganglia act to centralize the activities of receiving and reacting to stimulus from the environment. It is the centralization and cephalization (concentration at the animal's anterior end) of nerve cells that form the basic matter for the brain.

In more complex animals, vertebrates as dolphins are, the nervous system is actually comprised of two systems working together to create more complex behaviors. The peripheral nervous system consists of the sensory and motor neurons. The central nervous system consists of the brain, which processes the information gathered, and the spinal cord, which carries the nerve impulses to and from the brain. "The interplay between the central and peripheral systems allows an animal to sense environmental stimuli, integrate the information, respond appropriately, and in doing so, carry out the fascinating range of behaviors one sees in each animal and in the animal kingdom as a whole" (Postlethwait and Hopson).

Intelligence is usually related to animals with extended maturation periods, for the young with high levels of parental care. These animals tend to live a more social life with higher levels of intercommunication. Such is the case with dolphins.

Generally speaking, brain size is considered when measuring the intellectual abilities of a group or individual. Several different parameters are used such as brain weight to body weight, brain volume to body volume, and sizes of specific brain areas.

Scientists measure the percentage of body weight that the brain occupies across a variety of species as a comparison of brain sizes in different groups of animals. For example, the human brain is about

1.9 percent of the total weight of the human body as a group. Of course, that varies between individuals. In the Odontocetes (toothed whales), the percentage varies from 0.25 and 1.5 percent of total body weight.

Another expression of intelligence is the encephalazation quotient (EQ) or the ratio between brain volume and body surface area. The EQ of chimpanzees is 2.5, in bottlenose dolphins 5.6, and in humans 7.4 (Evans 1987).

The animal brain is divided into different areas each with its own responsibilities for processing information and operating different parts of the body. One such area is known as the neo-cortex, which is thought to be the area from which creativity and abstract thought is generated. The cortex is the outer, highly convoluted layers of the brain. In humans, the neo-cortex is ninety-six percent of the cortex and in bottlenose dolphins the neo-cortex is ninety-eight percent of the cortex.

It has long been known that dolphins possess a large brain, but the question still remains: what are they doing with those large brains? It is widely thought throughout the scientific community that the development of large brains in dolphins is due to their use of sonar imaging and vocal communication with other individuals and in social groups. Such complex behaviors and cognitive abilities would generate the micro-evolutionary pressures for the development of larger brains, but which came first—the behaviors or the large brains?

That question is easy to answer as I see it. When God created the Earth as we know it and populated it with animals and humans, He endowed each species with its own abilities and capabilities. Over time, species have changed and grown—each to their own abilities, some in stature, some in cognitive abilities, but each remaining within their own species and each growing from the basic abilities endowed by the creator, God.

All one has to do to understand what I have just said is to examine the two different models before us: the creation model shows us that at one specific moment in time fully developed, adult animals appeared on the Earth. We find this to be true when we examine the fossil record. In each example of the fossil record spanning through geologic time, species appear and disappear as whole, congruent examples of themselves. On the other hand, the model of evolution tells us that all life on this planet was derived from a single cell, which formed itself, by its own power, from a solution of inanimate material, which has not and cannot be proven by any scientific, empirical data.

So we see that ancient dolphins, although they may have had smaller brains than dolphins today, had the basic brain structure, which God has allowed to grow to its current size and abilities. Just as humans have not always been as intelligent as we are today with our televisions, automobiles, computers, airplanes, and rockets.

Over time, dolphins have grown in their cognitive abilities, just as humans have grown in their abilities to study dolphins. The fact that humans have only recently taken up the study of dolphins, and their cognitive abilities, does not mean dolphins only recently became "intelligent." To the contrary; I believe ancient dolphins were much more intelligent than ancient humans and it has taken this long for us to catch up to them.

How can I make such a statement? I have already demonstrated how ridiculous the model of evolution is. The model of creation is given to us in the book of Genesis in the Bible. In chapter 1, verses 20-23 we find that in the fifth day of creation "God created the great sea monsters, and every living creature that moves, with which the waters swarmed after their own kind, and every winged bird after its kind; and God saw that it was good" (NASB v. 21). In the sixth day of creation (vs. 24-31) "God created man in His own image, in the image of God He created him; male and female He created them" (v. 27).

Therefore, dolphins have been on the Earth for one day (or one thousand years) longer than "man." We have finally realized they are intelligent, but we still have not figured out the extent of their intelligence. Nonetheless, throughout their existence, dolphins have managed to survive in this world, generation after generation, as intelligent creatures, co-habitating with other creatures without destroying either themselves, other creatures, or polluting their own environment. On the other hand, humans may have been able to construct televisions, automobiles, computers, and go to the moon, but look at what we do to other humans and to the Earth itself as a result of our intelligence. I am in no way implying that "man" has the ability to degrade the planet more than God has the ability to restore it. We shoot and kill our neighbors, Children kill children in our schools, neglect and abuse our elderly citizens, abuse our children, and in spite of all the efforts to change our lifestyles we still live in a "throw-away" society. I prefer to think of dolphins as being higher on the intellectual scale than ourselves.

Of course, dolphins do not think the same way we do or at least not about the same things we do. Our concerns are not their concerns. It is only in the things that can't be put into words or described by any language that humans and dolphins can come together. We may not understand why a dolphin does not come rushing over to us when we call out to it. Dolphins probably can't understand why we ride around in boats. Furthermore, I doubt dolphins have any desire to enter our world in the air and on land, but when we enter their world common "grounds" may be found. In their world, they can learn more about us and faster than we can learn about them in our world. Is it any wonder why in most cases wild dolphins flee when a boat-load of humans jump into the water to be with dolphins. I believe dolphins can understand human intent within seconds simply by observing the human's actions.

I am sorry to say that we tend to think animals are not intelligent unless they can do what we can do. In other words, dolphins can live together with themselves and all the other creatures in the sea without having a murderous, thieving crime rate, but we do not think they are as intelligent as we are because they can't stick their heads out of the water and talk to us in the English language.

Dolphins can live just fine in their world without the aid of technology, but many humans can't even go into a restaurant without carrying at least one cellular phone in their pocket.

Dolphins use natural oceanic currents and tides and wind driven waves in their travels to preserve energy, while humans try to figure out which kind of energy sources we need to use after God gave us everything we need.

Throughout history, dolphins have dealt with their own renegade, radical or rebellious "colleagues" in their own dolphin way without psychiatrists, prisons, or mind altering drugs, but humans continue to consume record numbers of Valium, Prozac, and other psychological drugs while constantly searching for better and stronger drugs for mind manipulation that render the "user" to the order of a vegetable.

Only human arrogance says, "I will judge a dolphin's intelligence on their ability to mimic something I can create!" When we place one or two dolphins in a swimming pool and ask them to imitate a computer generated sound, how does that demonstrate the dolphin's intelligence? Why would some "out-of-the-blue" sound never before heard by the dolphin have any meaning to it? Just because we cannot understand the dolphin's vocalizations does not mean they do not have a language—they do! All one has to do is to watch a video of two or more dolphins swimming side by side and simultaneously they all change direction. How could that information be passed from one dolphin to the other as to when to change direction if it were not for some form of communication either through sound, thought, or gesture.

We humans want to stick our noses into everything we cannot understand. We want to change it, manipulate it so we can gain an understanding of it based on our standards and perhaps in doing so we overlook the most simple explanation for the reason we do not understand that which we seek. Most assuredly our greatest scientific minds are limited in ways of discovering the unknown. However, is it not possible for the old fisherman or the sailor who spends their life out on the sea to know more about dolphins and their behaviors than a scientist standing on the side of a concrete swimming pool glaring in at one or two dolphins that may have never actually been in the ocean before?

Here is the key; animals can think even as we do, but only sporadically and never in a consistent, enduring mode. In other words, a dolphin will always be a dolphin. Not to take anything away from them, I said they can think as we do, but I did not say we can think as they do and there is a difference! We may coax it and train it to engage our minds for a short period of time, but after that time it will go right back to being a dolphin, doing dolphin things, thinking dolphin thoughts. No matter how hard we try, we can never turn a dolphin into a human or visa-versa.

When my female Catahoula-Cowdog (Missy) was eighteen months

old, she showed me she was a smart girl, indeed. In my living room, I had a large mirror sitting on the floor, leaning against the wall. It was there for no other reason than I hadn't yet figured out where to hang it. Sitting at the kitchen table, I could barely see the top of my own head in the bottom of that mirror.

One morning as I was waking up with the usual cup of coffee, I noticed Missy standing in front of the mirror admiring herself. As a student of biology, my attention raised to a heightened peak because a simple tool used by animal behaviorists is a mirror. They want to see if animals can identify and distinguish between themselves and other objects as either being in the mirror or as a reflection being cast by the mirror.

The moment I noticed her interest in the mirror, I told myself to sit quietly so as not to disturb her, just watch and see what she does. At first, she just stood there looking at herself. Then she moved around as if to get a different angle. After a minute or so, she started looking around in the mirror at all the other reflections and then she started to paw at the mirror, a normal reaction. That is when I decided to intervene, before she lost her train of thought. I simply raised up my head to catch her attention. She stopped and stared at me in the mirror. When I raised my hand and waved at her, she turned and looked directly at me, then back in the mirror, then back at me directly.

I could see the gears turning in her little head, "What's going on here? He is sitting over there, but I see him sitting over here too."

This is where she demonstrated her intelligence. As she looked at me in the mirror, I raised my hand once again, but this time, instead of waving at her, I turned my hand over, making no sounds, and with my fingers made the "come here" gesture. Without hesitation, she turned and ran directly to me; to which, I greatly rewarded her with love.

Without planning it, without any design, by pure mishap and accident I was able to carry out one of the most important, basic experiments in animal behavior and animal intelligence sought after by graduate students and PhDs. around the world. In those few, short moments of excitement as I knew what was going on and wondered what the outcome would be, my understanding of animal intelligence unfolded before my eyes.

Sometimes, the most important lessons can be learned by just sitting and watching things as fragments of time pass by. If I had tried to set-up an experiment with the intention of getting Missy to do what she did on her own, there is no doubt in my mind that I would have failed. What I am saying is that in the process of planning and implementing experiments in controlled situations the very manifestation of the desired results can be lost before you even begin just by trying to force a particular situation to happen.

In those brief seconds of a million possibilities, God taught me a lesson that few scholars are afforded throughout their careers. In those few, short moments, Missy demonstrated without a doubt that she had figured out that she was on the outside of the mirror looking in at her own image. In fact, she knew the mirror was not three-dimensional as if there was an identical room behind a sheet of glass. She knew that as she saw my image in the mirror, I was not in the mirror, but sitting at the kitchen table. Had she not known, she would have tried to enter the mirror to get to me.

However, the most important lesson I learned from that event was not in the event by itself. Perhaps, the real lesson I learned came after the hugging and the petting and the loving. Not only did I find out that Missy can be, or is, intelligent, but after that she went right back to being a puppy dog playing with sticks, rubber balls, even chasing her own tail.

So how can we measure the intelligence of animals? Scientists spend hundreds of thousands of dollars and sometimes many years trying to pull off controlled experiments like the one I just described. After all that, they learn the same thing I learned; animals can and do demonstrate intelligence, or at least we can get them to do some things we want them to do. In the end, that intelligence is measured only by comparison to our own ability to manipulate the subject.

In the estuaries and the bays and the oceans of the world, dolphins are performing "the mirror experiment" day in and day out with no obvious concern whether a human notices or not. On a still day when the surface of the water is calm, from underneath looking up, the surface looks like a sheet of glass or a mirror when the sunlight hits it just right. Yet a baby dolphin learns within seconds after birth that the air it needs to survive is on the other side of the surface or visible plain. The dolphin is not perplexed by this, but a human on the other hand, can become mentally disoriented after spending time in the water and unable to visually determine which way is up. In scuba classes around the world, students are taught that if they become disoriented in the water, release some air and follow the bubbles to the surface.

Nonetheless, humans have an innate desire to stick their noses into anything they can think of to try to figure out, in human terms, how that "thing" works. Scientists write their proposals to get free grant money from governments and foundations to hire people and boats to go out and capture several dolphins, taking them away from life as they know it, place them in a concrete swimming pool, eliminate the less intelligent individuals, and begin the experiments.

Working with dolphins is a huge investment. The better scientists provide excellent medical care for their specimens along with all the artificial sea water needed to fill the pool and enough dead fish to fill their bellies.

There are several scientists worth noting here who have taken giant steps towards measuring the intelligence of dolphins. At the University of Hawaii, Louis Herman has shown that dolphins understand the meaning of words in their own languages and how word order affects meaning:

"Each of two bottlenose dolphins was tutored in a specially-constructed artificial language. One dolphin was specialized in a language in which words were represented by computer-generated sounds. For the second dolphin, words were represented by gestures of a signer's arms and hands only, as in human sign languages. In each case, the words referred to objects, actions, properties, and relationships, For example, the sentence LEFT BALL TOSS asks the dolphin to toss the ball on her left in the air." Or, "PERSON SURFBOARD FETCH, instructs a dolphin to take a surfboard to a person (who is in the water)" (Herman and Samuels 1990).

The dolphins were also asked to give YES/NO answers to spatial questions like HOOP QUESTION, meaning is the hoop in the water? For a deeper understanding of the study, you should definitely read the article; put your thinking cap on and get out your dictionary! The CONCLUSION of this article says:

"The message of these findings is twofold: (1) animals, in general, seem engineered primarily as efficient, broadband monitors of their world and through their genetic, developmental, experiential, and social endowments are able to acquire, retain, and utilize extensive knowledge of that world; and (2) the potential of animals for displaying language competencies is more closely approximated by examining receptive skills rather than productive abilities, bearing in mind that there are apparent substantial asymmetries in receptive and productive mechanisms, processes, and skills"(1990).

Also in Hawaii, Ken Martin has demonstrated dolphins' ability in abstract thought using television cameras and monitors (Martin, Ken. Nature. The Nature Series; The Nature Conservancy. THIRTEEN/WNET-TV; 1992). Dolphins in tank B watch television images of a trainer feeding dolphins in tank A. As the trainer leaves tank A, the dolphins in tank B leave the television monitor to meet the trainer and receive their food. This response shows that the dolphins understand the actions on the monitor are happening somewhere else. They are distinguishing between reality and its representation (abstract thought).

In Vallejo, California, Diana Reiss (Nature. The Nature Series; The Nature Conservancy. THIRTEEN/WNET-TV; 1992) uses an alternative method which is opposite (if you will) from that of Louis Herman, in that she has designed an electronic keyboard which allows the dolphins to issue commands to the trainers. If a dolphin wants to play with a ball or get a fish all it has to do is to push a specific key with its beak, a computer records it and produces a sound specific to that symbol and the dolphin gets what it asked for.

These kinds of experiments and studies go as far as to demonstrate particular abilities some dolphins posses. That does not mean every dolphin would respond in a similar manner when put to the test. In each case, one has to remember that certain animals were selected for certain tasks and what it comes down to is that an animal was taken out of its world and brought into the world of the scientist and trained over a period of perhaps years in order to complete the study. Furthermore, in each case, you see how the "human" has structured the experiment so they control complete manipulation of the subjects trying to get it to perform a specific task, which the human has designed with particular results desired.

Some humans pass algebra. Others fail. Can a determination of all humans be gained from that prospective?

Certain dogs have an innate sense of retrieval. However, rarely does a dog go to its master (in the beginning, the first time) and inform them that "I want to play catch with you." Rather, it is a case of the master recognizing the dog's abilities and expanding those abilities to the point where the dog will cooperate and follow certain commands.

Do not think that one day humans and dolphins will carry on a conversation in the ocean; dogs will be dogs, and dolphins will be dolphins no matter how long the planet exists. A wild dolphin out in the ocean hasn't the foggiest notion what is going on in the laboratory of a scientist in wherever. A wild dolphin could care less what humans are doing on the streets of Austin, Texas or in New York City. Even if a dolphin trained at "The University of Dolphin Training" gets out and joins a pod of wild dolphins, it will not teach them how to communicate with the professor at "Dolphin-U!" Don't be foolish, people; a trained dolphin will go right back to doing dolphin things at the first opportunity it is given.

That is why it is ridiculous to think we cannot release a dolphin born in captivity into the open ocean. I say it again, a wild animal is innately a wild animal regardless of where it is born. It does not matter how many generations pass—a wild animal is a wild animal! Their outward behaviors may change to fit their conditions, but their nature remains the same. Should a wild dolphin live its life in a concrete swimming pool?

7 IF SOUNDS COULD KILL

There must be a reason for the natural smile on a dolphin's face. I have thought about dolphins for some time now and have decided that their smile must be attributed to their uninhibited pleasure of life itself. Surely I jest. The dolphins' niche is not bound by fences or limited to mechanical transportation on highways that lead only to one place. They may never engage the beauty and majesty of terrestrial splendor away from the seashore, but they have the wonders of the oceans, which cover seventy-one percent of the Earth's surface with all the coral reefs, colorful fishes, kelp forests, and the freedom to travel at their own will. They don't have to punch a time-clock or wait for a paycheck or stand in a soup line for food. According to God's provision of food, I think a full stomach is left to their will alone, whenever and wherever they please.

More intriguing though, than a dolphin's ability to find food with sound, is an exciting question: do they actually use sound to debilitate their prey?

Studying a dolphin's intelligence is quite a different thing than studying its biological function and anatomy. For example, if a dolphin sticks its head out of the water in front of a group of scientists and claps its jaws together several times then disappears one scientists may say, "It's trying to tell us something;" another may say, "It's mad at us;" yet another may say, "There was no meaning to the jaw clapping." There is a wide variance in interpretation, wouldn't you agree? On the other hand, when surgical procedures reveal the only passage way for air to reach the lungs of a dolphin is through the blow-hole and not through the mouth there must be a consensus that dolphins do not breathe through their mouths, but only through their blow-holes—there is no room for variance in interpretation.

We are intrigued by dolphins partly because of their illusive nature. If you venture out into salt water for the purpose of finding a dolphin in the wild, you may not be successful. But that doesn't mean you didn't pass a dolphin along the way or that a dolphin did not know you were in the vicinity.

Dolphins can stay underwater for about seven minutes. Although they have eyesight comparable to our own, they obtain more information with their biological sonar than with their eyes. Since they have no obvious need to contact humans, they can stay underwater, fully aware of what is going on around them, and let you go right by without your knowing they are even there. Remember, it was the biological sonar of dolphins that sparked the Navy's interest to start handing out big bucks for research.

When you look at the physical structure of several different Odontocetes, you may begin to wonder how those animals catch their food. For example, the sperm whale (*Physeter sp.*) has teeth only in the lower jaw, no upper teeth, and the jaw itself is receded posteriorly from the anterior end of the head. Moreover, ". . . the massive structures of its forehead, which may involve thirty percent or more of its length" (Mathews 1938) "are now implicated in sound emitions" (Norris and Harvey 1972).

Sperm whales feed primarily on squid, but also an occasional octopus and a variety of fish, including salmon, rockfish, lingcod, and skates, all of which can swim much faster than the whales. At physical maturity, males reach a length of fifteen to sixteen meters and weigh up to forty-three and a half tons. How is it possible for the whales to catch such fast swimming prey, if not with the aid of stunning sounds? Whalers have seen live squid swimming out of the mouths of harpooned whales and when the whales are cut open, the squid inside the whales seldom show any teeth marks.

This, by itself, is not proof that any Cetacean has the ability to "stun" or kill its prey with sound, but it certainly does provide evidence of the possibility. You have a gigantic animal gliding through the water, searching the depths with its biological sonar. Surely it can create a burst of propulsion with the swish of its tail flukes and probably can swim at an accelerated speed for a short distance, but can it chase down and catch an animal smaller than the size of its own eyeball when that smaller animal can change headings in the wink of an eye?

Consider once again the location and configuration of a sperm whale's mouth. While it is large with respect to the overall size of the animal, it is small when compared to the size of its own head. With it located on the ventral side and receded from the front end, I doubt the whale can either visually see its prey in the direct vicinity of its mouth or if the prey would be in the field of its echolocation clicks. Added to that is the assertion that sperm whales sometimes feed at depths of a mile or more where vision is impossible.

The point I'm trying to make is that you have this giant animal swimming about in the dark depths of the oceans feeding usually on very small prey when as it approaches the prey it can detect it while the prey is in front of it, but at some point in the pursuit the small prey must be in a position where, if it was able to, it could evade the trap of the whale's jaw.

However, if the prey had been disoriented by a powerful sound produced by the whale and was unable to flee its capture then the whale could maintain its pursuit knowing where the prey was as it left the field of echolocation and knowing, of course, where its own mouth is located and successfully eat that prey. If you think about that for a moment, it makes sense. Close your eyes and touch the tip of your nose with a finger; even if you move your head after you close your eyes, you can still find your nose.

Before we look at the mechanisms for the production of sound in

Cetaceans and how much sound they produce, let's consider another Odontocete and its special circumstances, the narwhal (*Monodon monoceros*).

Adult male narwhals reach a body length of about five meters and weigh about sixteen hundred kilograms. The narwhal's long, narrow mouth is toothless; that is, as far as useful teeth go as we think of them. They have two teeth embedded in the upper jaw, which remain embedded in the females. However in the males, the left tooth erupts and extends forward as a tusk with a left-hand spiral. The narwhal tusk can grow more than three meters in length, with a basal circumference of about twenty centimeters. Some tusks can weigh up to ten kilograms.

So once again we have a large animal swimming through the ocean depths. Not as large and cumbersome as a sperm whale, but this one has a long tusk out in front of it that it has to compensate for in respect to moving through the water. No one is absolutely sure of the purpose of the tusk, whether purely ornamental or serving the purpose of attracting females for mating. I know of no instances where a narwhal has been seen with any kind of animal impaled on its tusk. Perhaps, the most useful purpose for the tusk would be digging up crustaceans from the bottom. Nonetheless, that kind of structure would produce an extra drag when trying to turn left, right, up, or down quickly in the process of catching its prey.

Odontocetes with many teeth or only two teeth embedded in the gums don't use their teeth for the purpose of chewing their food;

rather, just for catching the prey and holding it to position it for swallowing. Odontocetes swallow their prey whole—head first.

If indeed Odontocetes do have the power to stun their prey with sound, what protects them from their own sounds and from that of other whales? If a small fish were to be stunned by sound, it would manifest as a disorientation of the fish, an inability to know which way to flee, a fish that would be slowed in progress or stopped dead in its tracks.

If you know anyone who has suffered from equilibrium or vestibular disturbance, you know what I mean. Such afflictions are caused by or related to a malfunction or damage to the inner ear, which oddly enough is responsible for our balance and uprightness.

Sound waves travel four times faster in water than in air. This can be demonstrated (under safe, supervised conditions) by exploding a firecracker one to two inches away from the side of a glass bottle, both in the are and in the water.; In the air, the explosion will have no affect on the bottle, but in the water the bottle will burst.

Furthermore, the way we can detect the direction of a sound we hear is by the difference in time it takes to reach each ear, in the air, that is. But underwater, with the speed at which sound travels therein, we are hard pressed to give a definitive direction from whence the sound came.

The Odontocetes external ear is a little more than the size of a pinhole. "The melon, a waxy lens-shaped body in the forehead, focuses the sounds produced in the nasal passages so that they may be emitted in a narrow stream [forward] whilst returning sound waves are channeled through oil-filled sinuses in the lower jaw to the inner ear. The channeling of sound is made more precise by isolation of the inner ear from the skull by means of a bubbly foam. This reduces the interference from extraneous resonance" (Evans 1987, see Figure 5).

Figure 1.6 Sound propagation and reception in dolphins (adapted from Norris and Harvey 1974). Sound (both clicks and whistles) is probably produced in the region of the nasal plugs, refracted in some way (not yet fully known) to form a beam by the melon and, in the case of sonar clicks, the echoes received through the oil in the lower jaw, to the ear bulla

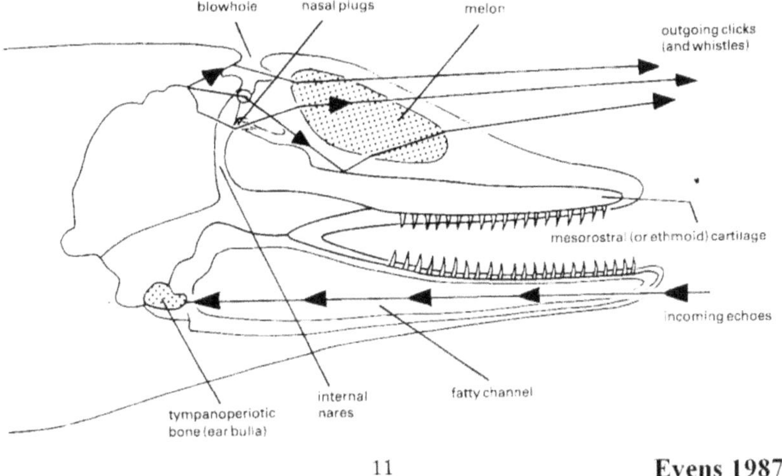

Figure 5

Ken Norris (1964) has written a "so called mandibular hypothesis emphasizing the key role of the lower jaw with its peculiar fat body as a specific sound conducting pathway." "The hypothesis postulates the presence of an 'acoustic window' on the lower jaw, which allows for the perception of sounds" (Popov and Supin 1990). With this protection of the inner ear against self-made sounds, how intense are the sounds produced by Odontocetes?

The environment in which Odontocetes live is dominated by sound sensations. Depending on water clarity and the depth at which the animal is located in the water column, vision is usually confined to the surface area—the interface between water and air. Therefore, God has endowed these wonderful creatures with an amazing biological 'sonar' system for communication and echolocation.

Odontocetes emit a variety of clicks, whistles, and squeals. Each sound has its own purpose and meaning to both the individual and their counterparts. Just like the different languages spoken by humans around the world, dolphins on opposite sides of the planet probably have different dialects in their sounds. While some sounds may be "universal," others may pertain only to a particular region or location. Nevertheless, dolphins must learn how to produce and interpret their sounds in the same manner we learn how to talk as a child—they learn from the adults they associate with as they grow and mature.

There is no doubt that dolphins are intelligent creatures. "Evidence for vocal learning is stronger among marine mammals than for most other mammalian groups" (Sayigh et al 1990). This makes perfect sense given the nature of their gregarious lifestyles, the environment in which they live, and the size of their brains. Caldwell and Caldwell (1979) say "that the first stereotyped sound produced by a young dolphin is the signature whistle." A signature whistle is the animals identification, much like our name. It distinguishes the individual in the social order and hierarchy of a pod or a group.

Unlike our names, however, each whistle is unique in one way or another.

Echolocation clicks are not emitted equally from all parts of the head, but are generally projected forward in a beam called the sonar field. Dawson and Thorpe (1990) conclude, "One intriguing possibility is that listening dolphins may be able to gather information from other dolphin's sonar echoes. Given their ability to process their own signals we would be surprised if they could not interpret (at least to some extent) the echoes from another individual's use of sonar."

Dawson and Thorpe stipulate, though, that this should not be interpreted as "language." I disagree! Noam Chomsky once argued with B.F. Skinner (the founder of operant conditioning) that animals were incapable of language and that such a concept would be anthropomorphic. I agree with Skinner in that, language is a form of communication and animals do communicate with each other, even other species. Would you not agree that "signing" is a language; after all, it is called "sign language." Chimpanzees and dolphins quickly learn gestures and hand signals from trainers, circus animals do as well. Therefore, they do posses language abilities, not in words as we know them, however. I do believe dolphins have and use their own language—Delphinese.

Dolphins can emit more than one sound at the same time and the signals can be produced in both a broad and narrow sonar field at the same time. These abilities, added to many other conditions, make it difficult to study sounds emitted by dolphins in the wild. Since dolphins spend most of their time below the surface it is difficult to know exactly where they are until they surface. "Obtaining meaningful measure of sound intensity requires knowledge of position and orientation of the animal in relation to the recording hydrophone, a suitable acoustic environment, and calibrated recording and analyzing instrumentation with adequate bandwidth and dynamic range" (Norris and Mohl 1983).

Believe me, that kind of equipment is expensive and difficult to find. Furthermore, getting dolphins in that kind of a situation is even more difficult.

Because of the difficulty in conditions of trying to study dolphin sounds in the wild, there are very few scientists who attempt the task. Scientists tend to not have enough patience or funding for that type of study. I admire those who do. Most scientists study dolphins in captivity.

The differences between studying dolphin sounds in the open sea or in a concrete swimming pool are vast and each has their pros and cons. In the wild you have a more natural situation, but confining a wild animal and obtaining the level of cooperation needed is very difficult. On the other hand, a captivating concrete pool provides a more suitable environment for the scientist, but the flat, hard surfaces of concrete create an obstacle for the animal in producing maximum sound levels sought by the scientist; those types of surfaces supply too strong of a reflection of sound back to the animal and the animal has no reason to produce a stronger signal. It would be similar to us standing in an echo chamber fitted with a public address (PA) system, the louder the PA system is turned up, the softer we would talk. Otherwise, the reflection of the sound bombarding our ears would become deafening.

"In captivity most Odontocetes produce relatively low-level sounds in the 140-180 dB//1 micron Pa range" (Diercks 1972). "In an experimental situation, however, approaching open water conditions, Murchison (1980) trained two *Tursiops* to perform sonar detection tasks against high background noise at distances up to seventy-three meters, using 2.54 centimeter and 7.6 centimeter spheres as targets. Under such conditions average source levels for single click series as high as 228.6 dB // 1 micron Pa / 1 yard were measured. This is five orders above the generally quoted levels" (Norris and Mohl 1983).

In order to grasp the significance of these figures, we must look at other studies which establish lethal sound pressures to small oceanic fishes. One such study was conducted in 1952 by Hubbs and Rechnitzer in which the following figures were established: for explosives such as dynamite and TNT, the lethal thresholds are between forty and seventy psi (229 and 234 dB // 1 micron Pa). (That is just .4 dB // 1 micron Pa above the levels found by Murchison in the study mentioned above). "The observed signs are hemorrhage and rupture of viscera, swim bladders, kidneys, or gonads" (1952).

These experimental results and observations are relevant to the problem of dolphin sound debilitation for three reasons: "First, peak levels equal to established lethal thresholds of fish can be produced by *Tursiops*; second, such levels are not harmful to their producer; and third, Odontocetes source levels measured in captivity or at sea generally are not indicative of maximum levels" (Norris and Mohl 1983). Dr. Norris also says the obvious experiment of training a dolphin to debilitate a fish with high intensity sound still remains to be done successfully (1983).

Do not jump to conclusions or be misled by this kind of science. While I am sure the experiments I've mentioned are valid, that does not mean that every single dolphin or whale in the world has those particular capabilities of producing sounds with such intensity; nor does it mean those particular animals produce that intensity of sound on an ongoing basis in the wild or that they could not have produced even higher intensities than they did during that experiment. What it does mean, however, is that there is at least a possibility that Odontocetes probably do have the wherewithal needed to debilitate their prey at their disposal.

If you or I could go out into a bay or other body of water, see and hear a dolphin debilitate a fish with sound, there would be no question of whether or not they had that ability, but dolphin feeding behaviors are difficult to observe in the wild. Even if there is a human that has actually seen and heard a dolphin debilitate a fish with sound, telling a scientist about the event would be pointless and meaningless because scientists do not take the word of amateurs (they may listen and note the event for future investigation) or even other scientists unless the act can be replicated under experimental conditions. You see the dilemma.

So scientists must again resort to the captive situations, which usually do not produce conclusive results, but may enhance an hypothesis. The fact that Odontocetes' prey are usually much smaller and faster than the predator gives rise to the hypothesis that Odontocetes must have some sort of aid in capturing their prey.

Seeking the answer to this question, Norris and Dohl (1980) experimented with three Hawaiian spinner dolphins *(Stenella longirostris*) using a dozen akule (*Trachuropsis crumerophthalmus*) as their prey. (It should be noted that akule are not normal prey for dolphins, and I will not speculate here why they were used). Twelve more akule were used as a control group in a separate tank with predatory jacks (*Cranx*). When the akule were placed in the tank with the dolphins, the dolphins immediately began working the pod of fish with "click trains." "They did so in a quite stereotyped manner; they always approached the school from its side, racing at it while emitting a whining crescendo of clicks. The fish school typically parted in front of the onrushing dolphins, both parts turning tightly towards the passing dolphin's tail. The mammal was unable to turn so tightly and had to make a broad circle before coming in for another run on the now reformed fish school" (Norris and Mohl 1983).

Although the dolphins failed to emit sounds strong enough to damage the fish, the polarity of the fish school was disrupted and split into smaller groups. One fish was so disoriented it was sucked down the drain of the pool. No akule, however, were eaten by the dolphins who continued taking their normal ration of freshly frozen fish. The control group of akule maintained the polarity of their school against the predatory jacks as a means of defense against predation.

In his Master's thesis at The University of Texas at Austin, Mchugh (1989) described seven different feeding behaviors of bottlenose dolphins he observed near Aransas Pass, Texas. Of the seven, one of those behaviors bears close resemblance to the observations by Norris and Dohl mentioned above. Mchugh observed "feeding rushes," which consisted of dolphins swimming rapidly near the surface for up to twenty meters and ending in a sharp "pin-wheeling" turn. These "feeding rushes usually involved two or more dolphins, and fish were often seen jumping in front of the dolphins" (1989). This feeding pattern usually took place in about three meters of water and was observed two hundred and forty-four times (26.0% of the feeding observations) during the study.

There are several observations from the field, which were not part of any experiment or formal study that need mentioning here. In 1963 and 1980, while on board an oceanarium's capture vessel, Ken Norris observed false killer whales (*Pseudorca crassidens*) feeding on several mahi-mahi (*Coryphaena hippurus*) at the surface (1983). As one whale held a fish in its mouth, two other whales tore it apart while another fish lay motionless on the surface apparently un-bitten. Only speculation could explain why the un-bitten fish did not flee; was it dead, stunned, or playing "possum," no one knows.

In the summer of 1973, Donald White observed a pod of killer whales (*Orcinus orca*) feeding on salmon (Norris and Mohl 1983). Off the side of his boat, he noticed a single salmon swimming rapidly away from the whales. Watching closer, he saw the salmon suddenly stop and lay motionless in the water as a whale smoothly scooped it up without changing direction or breaking stride.

Another incident was reported by Hardy Jones, a noted movie and documentary producer, after diving in the Bahamas. He observed a single spotted dolphin (*Stenella sp.*) accompanied by some mackerel-like fishes approaching a small, pearly Knife fish. When the dolphin was about a meter from the knife fish it fluttered up from the bottom. The dolphin made no further moves towards the fish. One of the fish accompanying the dolphin, however, moved forward and took the immobilized knife fish (Norris and Mohl 1983).

Whether or not dolphins stun their prey with sound is yet to be proven scientifically. The evidence gathered thus far, however, does suggest this to be true. While I have the utmost admiration and respect for Ken Norris, I can think of at least three reasons why his 1980 experiment produced inconclusive results: 1) the acoustic conditions of the pool were not compatible or conducive for the production of stronger signals by the dolphins; 2) the normal diet of the captive dolphins (dead, frozen fish); 3) the use of fish as prey, which are not typical prey for spinner dolphins.

There is further evidence implicating strong sounds in the debilitation of fish; it has long been an unethical practice for people to use explosives or old crank-phones to stun or kill fish. For this reason, in its Handbook of State Regulations, the state of Texas lists this practice under the heading of "Illegal Methods Of Catching Fish."

8 DOLPHIN THERAPY

Dolphins do have an illusive nature, which draws some people to seek them out. You, most likely, can't whistle at a dolphin like at a dog and get it to come to you. If you happen to be near a dolphin in a boat and jump in the water, thinking you can swim with it, you will probably never see that dolphin again.

Dolphins, in general, are leery of humans; we are noisy, obnoxious, and arrogant by nature. But sometimes in the world dolphins become our friends or even our companions. These situations are rare, indeed, but they do happen. Several of these situations have taken place in the form of "dolphin therapy."

There are a few people in the United States who have permits from the National Marine Fisheries Service to maintain captive dolphins for the purpose of letting their clients swim with the dolphins; not to be confused with the "Swim-With" programs discussed in a later chapter.

When humans interact with other species, something remarkable takes place. That is why we have domesticated so many animals. Domestication happens when humans interfere with the normal breeding patterns of wildlife and through artificial selection try to create the kind of animal or plant they desire.

Not to be confused with natural selection, which is the mechanism Darwin based his theory of evolution on. With artificial selection, humans can make drastic changes within a very small selected group of a species in a short period of time. Hence, from the naturally wild species of canine—the wolf, we get dogs like the weenie-dog and the poodle; and from huge jungle cats (tigers and lions) we get the house cat.

I must stress the point here that dogs are still canines and cats are still felines. Even with artificial selection, we can't turn a fish into a snake or a snake into bird the way Darwin thought evolution could do.

Nonetheless, we have domesticated plants and animals because few, if any, wild species will have anything to do with humans. So we change the make-up of the animal, take out the "wild" in them, and incorporate them into our lifestyles. Most relationships between pet owners and their pets increase the attitude and outlook on life of the pet owner. Pets usually give the owner an unconditional love, something that is lacking in this cold cruel world we live in.

While dolphins are not domesticated (may it never happen), they are held in captive situations, which have shown to be beneficial for handicapped humans. It is not suggested that dolphins posses some hidden power to heal, but swimming with a dolphin can result in such an emotional uplifting for a human, which allows for a positive attitude, in the afflicted, and self-healing. For this reason, these conditions of captivity are the only ones I can agree with. However, not in a concrete swimming pool.

In 1978, Betsy Smith, an educational anthropologist, set out to investigate whether dolphins could help her penetrate the guarded worlds of seemingly impervious autistic children. "When you work with autistic children and adolescents, you need to pick up their subtle body cues to understand what they are doing and thinking. Dolphins are excellent at reading the body language of people, says Smith" (Livermore 1991). Her first project was 16-year-old Michael who had been diagnosed as non-verbal. "At the end of a year, 16 encounters, his teacher had noted that Michael was much happier, easier to work with, and his attention span had increased from a few minutes to half an hour. He had also stopped biting his fingers and smacking his head, overcame his fears of stairs and elevators, and was able to smile and show affection" (1991).

Living From The Heart is a non-profit organization in Parker, Colorado, which tends to terminal cancer patients. Stephen Jozsef has taken patients to the Dolphin Research Center in Grassy Key, Florida to swim with dolphins. "Dolphins, Jozsef thinks, exist in an alpha state a meditative condition associated with creativity, intuition, and perhaps the potential for self-healing" (Kaplan 1989).

At another institution (I'll not mention the name here because they also allow the public to swim with dolphins for a profit), Julie Baxter, a licensed occupational therapist, has worked with several kids who have cerebral palsy. Their therapy involves a rigorous schedule to train and maintain their muscles. Baxter says, "Both kids have made a lot of progress in a short amount of time. I'm not sure how, and there's no scientific data to back it up, but somehow these animals seem to have a way of identifying and responding to the needs of special people " (Livermore 1991).

At the Dolphin Research Center, David Nathanson, a clinical psychologist, is studying the effects dolphins have on retarded children. Using the dolphins as stimulus and positive re-enforcement, he has taught kids between the ages of two and six the alphabet and complete words in as little as six months. David Nathanson says, "These findings demonstrate conclusively that dolphins can enhance the attention span of mentally handicapped children and thus increase their rate of learning" (Livermore 1991).

The work of these scientists is invaluable to the research of human/dolphin communication; however, each has expressed that the data gathered from their work is merely speculative, not empirical. They do not know why dolphins have such a positive effect on people, just that they do. Perhaps we will never know the answer until we arrive in the Kingdom of God.

I think Ken Norris sums it up the best, "We're not on a pedestal someplace, alone and magnificent. Dolphins are one branch of a tree we are all part of. We are of them, they are of us, and the more we know about them and the other animals, the fewer the barriers there will be between us" (Kaplan 1989).

9 CONDUCTING RESEARCH

This chapter may not be for everyone, but then again, who knows? When I started college at the age of thirty-five, the only thing I knew about research was that it was something being done by a "geek" in a laboratory with test tubes and gas burners. I quickly found out that was not the case.

While the reasons or the motives vary, you are actually conducting research each time you ask someone a question:

"Hi. What's your name?"

"Joe."

"I'm Glenn. Are you from around here?"

"Well, not really. I'm from Kansas City, but I've been here for almost two years."

That's research. In a matter of a few seconds, I found out Joe's name, where he came from, and how long he's been around. That may have been just an everyday conversation on a street corner or at a bus stop, but then again, I could have been standing on a street corner poling people as they walked by to find a ratio between native and non-native citizens of a city or neighborhood.

The word "research" has so many meanings and mechanisms that it can be interpreted in almost any way. Research is simply the act of trying to find information to a question for which you have no answer. If you measure the length of a board, check the air pressure in a tire, look up the phone number for a book store, that's research. Research makes a scientist, not the other way around.

Research can be more narrowly defined if you add more elements to it: a specific intention in mind, specific parameters, method used for gathering information, and if needed, a control for comparisons, etc.

For my purposes here, though, all I wish to do is to open some pathways for literary research (the written word) and some basic field study. While these methods will work for any subject, I hope you'll use them to find out more about dolphins and whales.

It is very gratifying to watch something on a television show or hear a conversation about a subject in which you have personal knowledge. It gives you the opportunity to watch and listen for correct or incorrect statements. It allows you to participate in the exchange of information. It is also easier to learn more about a subject if you already have some familiarity to it. For example, when I see a segment on the news about some whales that are stranded on a beach, I now know what to watch and listen for: what kind of whales, how many, what social structure, what authority has charge over the situation (I may even recognize a name or a face), what is the prognosis, or do they have a clear plan, so on and so forth.

When you learn more about any one subject that opens the door to learn more about related subjects. As I learned more about dolphins, I educated myself in oceanography, meteorology, animal behavior, psychology, animal husbandry, ichthyology, and politics.

For the person who has just seen a television show or a movie or a news segment about dolphins and you want to begin to learn more about dolphins, perhaps the best starting place is the local library.

The good thing about research is that you can go at whatever pace you are comfortable with, depending on your zeal and level of interest. Libraries are a wonderful place, and I no longer feel like a nerd when I walk into one. At first, they can be a little intimidating, but don't let that scare you away. Librarians are trained in the art of gathering information. Reference Librarians usually have at least a Master's Degree, sometimes higher, and they do what they do because they love it.

As I was learning how to research in my college library, I began to feel like I may be bothering the librarian too much. Mind you, that was just my feeling, she had done nothing to make me feel like that.

One day I went to her with a heart-felt "Thank You" and she replied with the same saying, "That's quite alright, Glenn. If it weren't for people like you, I wouldn't have a job. I'm a question and answer person . . . when you ask me a question, if I don't have an answer, I'll do everything I can to help you find it."

Even if you think you know how to find your answer, take the time to say "Hello" to the librarian and tell them what your looking for; chances are, they probably know of another resource you had not thought of. Furthermore, it is not out of the question to call before you go to see if indeed they do have information on your subject. That gives you a quicker starting place when you get there.

Most libraries are computerized by now. Don't worry about computer literacy, you'll get the help you need to get started. Believe me, you learn by doing. Most library computers let you search by "subject, author, title," or in combinations. Feel free to play around a little and get familiar with the system.

Start with a simple "subject" search, like "dolphins." You'll be amazed at how many publications you will find with something in them about dolphins. Some computers search either "books" and others "periodicals," while others perform combined searches. It all depends on how your library is set up. So again, play around for a while and get used to the system.

Some computers only list the material contained in that library. Others may be linked to a network that also searches other libraries;

if so, it should tell you where that publication can be found. If you find something you want to read and it is not located at that library, ask the librarian if they will obtain it for you through an "Inter-Library Loan." There may be an option to see an "abstract" of the publication, these are invaluable tools for narrowing your search! Most libraries have a printer connected to the computer so you can print out the results of your search.

Books are a great source of information. They range from children's reading level through the collegiate level of reading. I know most people usually just read the text of a book, but I have found a great deal of information in the pages before and after the text. Many books offer appendices, glossaries, indices, and bibliographies. I find these to be most informative because they provide statistics, graphs, tables, explanations, and most of all, bibliographies provide you with an abundance of other resources to expand your research. Bingo!

Periodicals are also a great source of information, but you have to be careful here; not all magazines, shall I say, tell the whole truth and nothing but the truth. You will find that some magazines tend to sell their product through sensationalism. You will find that journals provide more reliable, scientific information complete with bibliographies and an abstract of the article's author's credentials.

A library just wouldn't be complete without a "Reference" section. It is that corner of the building or section that's always the quietest and you hardly ever see anyone in it. That's where they keep books like dictionaries and encyclopedias; the kind of books you would never read from cover to cover. But, look a little closer and you'll find some great research tools there. Once again, don't be afraid to ask the librarian to show you around.

Reference books are usually confined to the library, you can't check them out. But you can make notes or copies of what ever you want. If you want to find out what company or corporation does what with dolphins—there's a book for that. If you want names, addresses, or phone numbers—there's a book for that. If you want business financial statements and company structures—there's a book for that. You want information about laws, government agencies, or Congress - there's a book for that. You will be amazed at how much printed information you will find in the Reference section if you just do a little research.

If you're fortunate enough to live near a college or university, don't be afraid to visit one of their libraries, if they allow the public. It is a common misconception that they will not let you in unless you're a student there, not true! You may not be able to check-out books, but then again you may. Ask them if they issue temporary guest passes. If not, and you find something you would like to read, go back to your local library and ask for an Inter Library Loan. Either way, college libraries usually have many more resources than your local library. I'm sure they will not mind you using their search computers, taking notes, and making photocopies.

Finally, if you're the kind of person who likes to be outdoors going to the woods, *walking through the park*, fishing ,or hunting, I hope you will be inspired to do a little field research. It is not hard to do and it will enhance your pleasure each time you go out. God has created a wonderfully complex world for us to enjoy and a complex brain so we can gain understanding about it.

The greatest tools for learning about the things we enjoy outdoors are "field-guides." Field guides are illustrated books that help us identify things like wildflowers, trees, butterflies, birds, rocks, grasses, fishes, mammals, snakes, dolphins, etc. Of course, they all vary by author and publisher.

Field guides are illustrated either by drawings or photographs and provide a biological description of the species, even rocks are classified taxonomically.

Once again, you can find a variety of field guides in your local library. Eventually, you'll want to purchase your own so you can make notes on what you find. Some have what is called a "Life List," which lets you keep track of your findings. There are some bird watchers who travel hundreds or thousands of miles to add a single bird to their Life List. It is very rewarding to look back over time and remember where and how you saw a particular species.

As you begin to plan and structure your outings to enjoy wildlife, keep a journal. A spiral notebook works great for this. Make notes of your experiences: date, time of day, temperature, sky, and weather conditions. Describe what you have found or seen and the circumstances that led you there. By all means, draw pictures and don't worry about how funny they look, you'll get better each time you do it.

If you are the kind of person who likes to search the internet, you're on your own. Don't believe everything you see or read.

In my studies of dolphins, I was fortunate enough to live in Port Aransas, Texas while attending Corpus Christi State University (CCSU).

Port Aransas is located on the Texas gulf coast on a barrier island (Mustang Island). Corpus Christi Bay is behind Mustang Island and on the mainland side of that bay is where Corpus Christi is located. It is about a thirty mile drive from Port Aransas to Corpus Christi.

Mustang Island is thirty miles long (north to south) and about a quarter of a mile wide. Port Aransas is on the north end of the island and the only access to the north end is by ferry operated by the Texas Department Of Transportation. The south end has a bridge. The ferries are one of the attractions to the area. They cross the International Ship Channel where the world's shipping industry gains access to the Corpus Christi Port. Riding the ferries gives people a great opportunity to see wild, free ranging bottlenose dolphins (*Tursiops truncatus*).

That region of the Texas gulf coast (the coastal bend) is loaded with dolphins, both resident groups and transients. And dolphin research is what drew me to CCSU. Back at Austin Community

College (ACC) in Austin, my favorite professor had given me the nickname, "Dances With Dolphins," a name I got after many hours of research and love for the animals.

Dr. Estes is the greatest teacher I ever had. She didn't confine her classes to a classroom. We went on a lot of field trips to a lot of different places and one of those places we went to once a semester was the Marine Science Institute at Port Aransas. It is one thing to study books, magazines, and research papers, but if you really want to learn about the subjects on those pieces of paper you need to get outside and investigate for yourself.

On those field trips to Port Aransas, I spent as much time as I could sitting by the water with a pen and notebook, my binoculars, and my camera watching and studying dolphins. Although I consider myself to be "some what" of a scientist, I have always hated sounding like a scientist in my writing as demonstrated by the following notebook entry:

"These two dolphins are working the harbor as a team. Roxanne and I watched them come into the harbor, work the area for food, then leave. After watching them (about 20-25 minutes), I can imagine their strategy went something like this: "I'll take the right side, you take the left. We'll go around the edges to scare out the fishes hiding in between these moored boats, then we'll school them all together out in the middle and munch out!" One of them was more visible at the surface than the other, so they may have "said" this to each other: "You go down to the bottom and scare 'em up here, I'll herd 'em to the middle and we'll munch out."

On a particular Port Aransas field trip, a few brief, surprising moments were worth the whole trip:

"Last night, I walked out to the water at 10:00 PM. As I approached the junction of two jetties, I heard a distinct 'whoosh.' The sound was faint, but enough to catch my ear. I stopped dead in my tracks. I had heard that sound before—it was the sound of a dolphin blowing (breathing). I tried to become invisible so I could sneak up on him. In the 90 degree corner of the two jetties, I caught a glimpse of not one, but two dolphins still in the water. As I took another step, they saw me and fled the scene. Nonetheless, I saw them from about 15 feet away and my heart was beating fast. Since dolphins rarely stop still in the water, I think I walked up on them while they were sleeping and since dolphins sleep one half of their brain at a time that would explain their awareness of me walking up on them."

That kind of experience is priceless to a student. I was excited. God causes those kind of things to happen. The possibilities of other things I could have been doing are too numerable to count. And God gave me the experience not another student who probably would not have recognized the situation for what it was worth. Dolphins are difficult to study because of their natural habitat. Although they live in water, they are never too far from our natural habitat—the air.

In my second semester at CCSU, I took my favorite of all courses, or I should say I took a course on my favorite of all subjects—Marine Mammals. Imagine a whole course geared toward Cetaceans and other marine mammals. I was in dolphin heaven.

There was something very strange about that instructor and that class as well. Although he did a lot of work with dolphins, I could not see that he loved the animals, let's say, like I did (do). He was the "Head" of the Marine Mammal Stranding Network for the area. He had a bottlenose dolphin skull in his office (Since dolphins are protected under the Marine Mammal Protection Act Of 1972, you must have a permit to posses any part of a dolphin, including bones!), but he acted like it was nothing more than just a paper weight, no more important than the globe sitting in the corner. I had to try to convince him that I wanted to be a part of that network, a major part; perhaps I came on too strong. Maybe he did not want my services.

I think I was the only one in that class because of love of dolphins. Everyone else took the course as an elective. They were chemists, engineers, English majors, and scuba divers. I was there because if I could be anything else in the world, I would be a dolphin. I later found out that he was calling anyone else in the class to investigate stranded dolphins, but me. What an insult.

Half-way through the semester He called me at 10:00 PM: "We have a live stranding, we're gonna need some help. If you want to, come over to the State Aquarium as soon as possible." I thought wow, he's finally giving me a chance. I grabbed my keys, jumped in the car and was there in thirty minutes. I was the last one there. They had all been there since about 3:00 PM.

I remember standing at the tank, as if glued to the glass, with tears of awe running down my cheeks and praying to God: "Thank you God for letting this animal be found by humans before she died on the beach. And thank you for allowing humans to know what to do with her and letting me be a part of it. Dear God, please help this beautiful, amazing animal recover and go back to her home. Thank you God."

Nonetheless, I did manage to get in some actual working time with the dolphin later that night. The Texas State Aquarium was fortunate in being able to handle such a situation as far as equipment and medical care. They pumped sea water into a 15 x 8 (I don't remember the dimensions of the tank) steel and Plexiglas tank.

I'm not sure where from, but they called in a veterinarian; I have reason to believe he came from Sea World Of Texas (San Antonio). The aquarium provided their animal husbandry specialist for care management. Tony from the Marine Science Institute in Port Aransas and the Marine Mammal Stranding Network Of Texas were also involved. However, volunteers were needed to actually handle the dolphin in the tank.

For that first night, and the two to follow, I donned a wet suit and entered the water with the dolphin. She could no longer keep herself afloat or swim, so our job was to maintain her position at the top of the water so she could breathe, and walk her around the perimeter of the tank. It is amazing how light a four hundred and fifty pound dolphin is in the water. Although not under my chosen circumstances, my dream had come true. I was in the water with my hands on a dolphin actually holding her afloat so she could breathe and continue to live. What a blessing! All the time I had spent in libraries reading and researching, writing papers and letters, talking on the phone to dolphin people, and just thinking about dolphins was being paid back to me.

It was a very humbling situation. I couldn't stop myself from crying from time to time as I held her head out of the water and stroked her smooth streamline body. Imagine, there I was, with one other person, not only in the water with a live dolphin, but this beautiful creature who had lived its life swimming, playing, socializing, and eating food out in the open ocean was now depending on me to keep her from drowning. It really made me think of how God had created all the different animals in the world and how truly amazing this animal was. I tried to keep my mind focused on God as I talked to her: "Come on girl, get better, you can do it. You're such a beautiful girl, come on baby, get well. You're not alone, we're here with you. Come on girl, you can do it." The next morning, I entered this note in my journal:

"3-22-94 found on Padre Island. Female live stranded delphinid, species questionable. May be short-snouted spinner dolphin (*Stenella clymene*) or Fraser's dolphin (*Lagenodelphis hosei*). Blood profile shows liver damage. She has loss of muscle control (listing, twitching & quivering; central nervous system). Her eyes have been closed or slightly open. She tries to make the "swim motion," but without success. She is a beautiful animal about 2.7 m. I think she is a *clymene*."

Despite our efforts, she died four days later. But, several hours before she died, her colors brightened up and her color-patterns were finally distinguishable. It was determined she was a Fraser's dolphin.

However, our task of dolphin rescues was not over. The day after dolphin number one was rescued, a second dolphin was rescued from the Padre Island beach. This time it was a male bottlenose dolphin, and although he was disoriented, he was in better condition than the first:

"3-23-94 The second live stranding in as many days from Padre Island. A male (*Tursiops*) either (*truncatus* or *aduncus*) about 2.5 m, many teeth rakes on body. He seems to be healthy, swimming on his own & responding to up-rights by handlers. Enzyme count indicates heart trouble. He may have been beaten up & thrown out of pod."

He spent his first day at the aquarium isolated from the first dolphin until his condition was evaluated, then he was transferred to the larger tank with dolphin number one. We hoped it would lift the spirits of the first dolphin to have company of her own kind in the tank, but that was not the case.

This was a young dolphin and in spite of his condition, he had a feisty spirit. We named him, "D-2" (Double Trouble). The vet treated his open wounds and we fed him medicated formula through a tube, which was a task. But he quickly started accepting fish as food and we would medicate the dead fish before his feeding.

"3-24-94, 7:15 AM Tony is trying to feed him some small, live fish we caught with a seine net. He checks it out, but has not taken one yet."

He responded very well to the medication and soon needed no assistance from handlers. So well, in fact, plans were initiated for his release and a debate quickly ensued concerning his destination. Sea World Of Texas was lobbying hard to take him to San Antonio, but fortunately the real authorities determined he should be returned to his natural habitat, the Gulf Of Mexico. His release, though, did not take place before there was yet another live stranding.

By this time, the aquarium had moved D-2 to their annex building and volunteer management had changed hands so many times it was hard to acquire a time-slot to work with the dolphins. Moreover, the news of all the strandings attracted the media and the National Marine Fisheries Service (NMFS) who is charged with the responsibility of the protection of marine mammals.

At the time of his release, I had not had the opportunity to work with D-2 for almost two weeks. The stranding network was still in charge of his care and coordinated with the Marine Science Institute to use UT's research vessel the Longhorn for transporting D-2 out into the gulf. Fortunately for me, the Longhorn's home dock was three blocks from my house, and I could not pass up the opportunity to say goodbye to D-2.

I was not slated to board the Longhorn, but minutes before departure Tony (Figure 6) saw me standing on shore and called out to me, "Glenn, don't you want to come along . . . get on board!" It took about three and a half hours to reach our destination of about twenty-six miles out in the gulf. Along the way, a Cetacean researcher from Oregon State University, Bruce Mate, attached a satellite radio transmitter to D-2's dorsal fin so his where-a-bouts could be monitored (Figure 7 on pg. 114). Of course, I had my camera equipment with me and recorded the entire event on film. It was a wonderful experience to be a part of a team who rehabilitated a sick dolphin and took him back to his home.

Figure 6

Conducting research is like anything else in life, you get out of it what you put in to it. The further you indulge yourself seeking answers and information, the more you will feel rewarded and the smarter on the subject you will become.

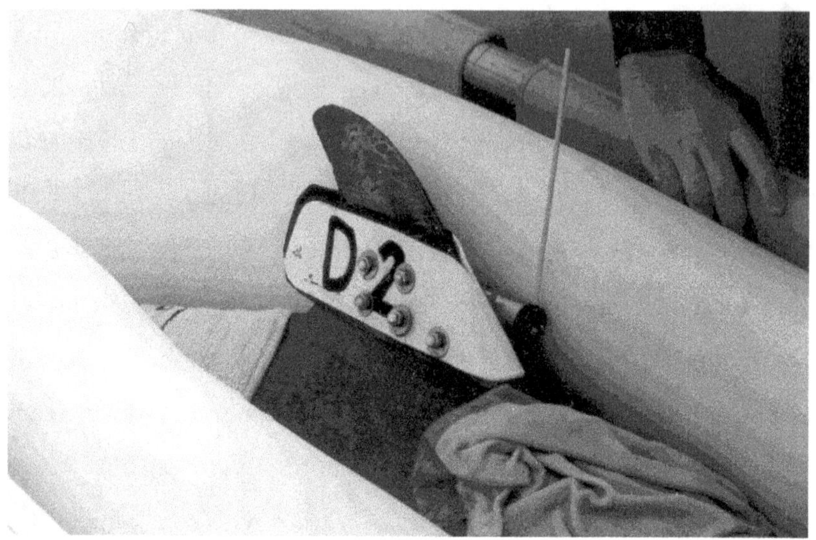

Figure 7

Don't confine yourself to the opinions or statements of just one person, including myself. Check what you hear and read against that which others have to say on the matter. Weigh the evidence in front of you and make your own conclusions. You have a brain, use it!

PART THREE: EXPLOITATION

Any time you abuse or misuse a living creature you are exploiting it's natural, God given traits and characteristics. Exploitation can take the form as results caused by a direct action against the victim(s) or as indirect results from actions taken elsewhere without regard for the consequences. Nevertheless, exploitation can be harmful to either a particular individual animal or to whole populations, depending on the location and the manner in which the actions are taken.

The exploitation of marine animals increases as the human population continues to grow exponentially. There are four particular areas of Cetacean and dolphin exploitation that cause me great concern: the needless, in-humane killing of dolphins and whales by the world's fishing industry for no reason other than stupidity; the reckless industrial pollution of the oceans; dolphins in captivity for entertainment (money); and the cultural differences around the world, which causes the slaughter and even the consumption of dolphins by humans.

Humans tend to manipulate every possible situation or thing that allows them to with only the end results in mind rather than pondering the means to an end. I wish I could say the exploitation of marine mammals was just an oversight or a lack of education, but sadly enough that doesn't seem to be the case.

10 THE FISHING INDUSTRY

The night was long and sleep had been spotty as the wind and the sea tossed the large vessel back and forth, but as dawn was breaking the sea began to calm. The captain was in the galley getting his first cup of coffee when the voice came over the radio on his side, "Captain . . . captain."

"This is the captain, go ahead."

"We have a pod of dolphins on the screen."

"Distance?"

"Three nautical miles, sir."

"Plot an intercept course and ready the chase boats. I'll be on the bridge in fifteen minutes."

The captain summoned his first-mate to the galley to discuss their plans for a successful catch; he wasn't about to lose this one. The first mate listened as the captain explained his strategy and then he went on his way to carry out the captain's orders.

On the bridge, the captain checked the status of his vessel, "Are we on course with the pod?"

"Aye, captain!"

"Is the engine room ready?"

"Aye, captain, ready and standing by!"

"Are the chase boats ready?"

"Aye, captain. The chase boats are ready . . . net and deck crew all ready and standing by, sir!"

The captain gave the order, "Helm, full ahead! When we have visual confirmation of the pod, veer starboard and parallel their course and slow to one-quarter to release the boats, understood?"

"Aye, captain!"

The large vessel gained speed as it cut a wake in the open ocean. Excitement grew as adrenaline began to flow through the veins of all the men on board. It was time to go to work. Everyone knew they all had to work together for a successful catch. This is the moment for which they went to sea.

It sounds like that ship was out to catch a pod of dolphins. Indirectly, you could say they were, but what they really wanted was underneath the dolphins—under the surface of the water. Years ago, in the 1920's, the world's tuna fishing fleets discovered that yellowfin tuna congregate and travel beneath pods of dolphins in the Eastern Tropical Pacific (ETP) Ocean.

Yellowfin tuna is the most popular type of tuna you see on the store shelves. You find it in the cans labeled Chunk Light Tuna. The other types of tuna you find on the shelves, like Fancy Albacore or White tuna, are not caught in association with dolphins. The association of yellowfin tuna and dolphins only occurs in the ETP and usually involves spinner, spotted, or common dolphins in herds of up to thousands of animals.

Back in the 1920's, the tuna were caught without harming the dolphins by throwing bait overboard. The tuna would go into a feeding frenzy near the surface making them easier to catch and the dolphins would simply swim away.

In the 1950's, however, the demand for tuna was increasing with the world population and pulling in tuna one-by-one no longer proved efficient. With the advent of larger, better equipped vessels, the fishermen found that if they catch the dolphins they catch the tuna. The more tuna they caught at one time, the more money they get quicker. Human nature—greed!

The method used to catch the dolphins and the tuna is called "purse seining." "The modern purse seiner can cost six million dollars. It's two hundred feet long, carries a crew of 16, and, using advanced satellite navigation, can cruise at fifteen knots for three months. On a single trip it can catch and freeze a thousand tons of tuna - worth $800 a ton, and it makes three or four trips a year" (Linehan 1979).

So what does all this have to do with exploiting dolphins? Here's what happens: the tuna swim with dolphins and the fishermen search for dolphins. As they approach a pod of dolphins, smaller boats, sometimes a helicopter, are launched to herd the dolphins and deploy a net around the herd. This may or may not be an easy task. Sometimes man is smarter than dolphin; other times, dolphin is smarter than man. Humans, in the form of tuna fishermen, can't stand to be outsmarted by animals so they add a little fire-power to their tuna catching weaponry; they throw small bombs in the water around the dolphins to disorient them. Remember our discussion in chapter 6 about bursts of sound in water. Since sound travels four times faster in water than in air, water amplifies the waves of concussion from any explosion.

It stands to reason that if dolphins use sound to debilitate their prey, then if the sound is increased, at some point dolphins would also be affected. The bombs thrown in the water by tuna fishermen probably have the affect of deafening the dolphins and interrupting their communication and locating abilities. Furthermore, this may disable their echolocation capabilities, rendering the dolphins blind in murky water. Dolphins have been observed diving quickly and deeply to escape the bombs only to become entangled in the net a hundred feet or more below the surface and consequentially drowning in their attempt to get away. And how close to a dolphin would an explosion have to be to actually kill it on detonation?

A blind dolphin can survive in the ocean, but a deaf dolphin is as good as dead. While dolphins rely mainly on echolocation to gather information about their environment they still posses very sensitive hearing capabilities; a bomb explosion in the vicinity of a dolphin could bust its inner ear.

Then as if bombs weren't enough to ruin a dolphin's day, they have to deal with a giant net the tuna fishermen have deployed around them. First of all, it is uncertain whether or not the nets are detected by the dolphin's biological sonar. This is speculated when the net is hauled in, by the numbers of dolphins found entangled in the net—drowned. Either they can't detect it or they are so disoriented by the bombs they don't know where they were going. Underwater, a net has a "clinging" property; that is, it wraps around whatever comes in contact with it.

You may ask, "why don't the dolphins just jump over the net to safety?" If the dolphins can detect the net with echolocation, they cannot see what is on the other side of the net. Close your eyes and imagine you are crawling along the ground in complete darkness and you arrive at what you conclude to be a cliff. You feel the ground fall away from you, but you can't feel the bottom with your fingertips. You search for something to throw over the side, but find nothing. Not knowing how deep the cliff actually is, would you stay where you are or go over the side? The cliff could be three feet or three hundred feet. Remember, dolphins are intelligent and sentient, they do not want to commit suicide. So they remain inside the net scared for their lives, uncertain of their fate. And they die.

With all that said, the net is deployed around the dolphins, perhaps a mile in circumference and a hundred feet or more in depth. Once that is completed, the mother ship begins drawing the bottom of the net together—closed, which eliminates any chance for escape by the dolphins on their own initiative. This process is the "pursing" in "purse seining."

When the pursing is complete, the mother ship begins pulling the catch on board using a vast array of heavy mechanical devices. As the net is drawn together, the catch is concentrated by the decreasing area within the net. This is when many dolphins drown. Should they survive that point in the process, they will probably be crushed by the equipment as they are hauled on board due to their physical size being much greater than the target species—the tuna fish.

Now, let's suppose a dolphin survives the whole ordeal and ends up on the deck of this large vessel. What happens then? There are numerous possibilities, all of which are detrimental to the dolphin. You first have to realize that after enduring such traumatic circumstances they probably have some form of physical injuries, either external or internal, even emotional. I've seen too many (one is too many) pictures and videos of dolphins on the decks of ships with their fins or tail flukes ripped of their bodies and blood gushing out of the open wounds. Or the dolphins are in tact (external anatomy), flopping around the deck and a deck hand is walking around beating them to death with a baseball bat. Either way, they are then kicked or hurled over board only to fall thirty or forty feet to the surface of the water. How long has it been since you did a thirty foot belly-flop?

Now tell me, is that not the exploitation of dolphins? The dolphins help the tuna fishermen find their prey, then the fishermen cold-heartedly and intentionally slaughter the dolphins. Nothing in the world makes my blood boil more than that! Well, almost nothing.

Now it's time for the gruesome statistics, the mortality rates suffered by dolphin populations in the ETP at the hands of the tuna fishing industry. But first, let me say this information is only estimated and actually unreliable, because after all, who, other than God, has seen and counted the death of every dolphin killed in a tuna net or at the whim of a deck-hand on a tuna boat?

Furthermore, no one wants to admit to the true numbers. The United States Congress charged the National Marine Fisheries Service (NMFS) with the authority of the Marine Mammal Protection Act (MMPA) of 1972 and they want the mortality rates to comply with the MMPA so they will appear to care about the wildlife and the environment so they can get re-elected.

The NMFS wants the numbers to look good so they don't get fired or have to spend the day in front of a congressional committee.

The tuna fishing industry wants the numbers to look good so they don't get shut down by an angry public.

The animal rights groups and conservationists want the numbers to look bad so they will still have a job and gullible people will continue to throw money at them. After all, if no one is killing marine mammals, why do we need a Center For Marine Conservation? All this in the name of a tuna fish sandwich.

With all that said, here are some numbers just to give you some idea of what has happened to dolphin populations by the tuna fishing industry:

- From 1960 - 1990 6 million dolphins killed.

- 1975, 166,645 dolphins killed by U.S. fleet alone.

- After the MMPA (1972 - 1990) which calls for a 0 kill quota, 800,000 dolphins have been killed.

- "More than three hundred dolphins were killed every day, twenty-three per hour! An average of one hundred thousand dolphins have been killed every year by the tuna industry in two decades" (Earth Island Institute. Dolphin Alert. Spring/Summer 1990).

Most of the above numbers are related to the U.S. tuna fleet only. While they were probably the largest participant in the tuna fishing activity, other countries also played, and continue to play, a big part in the senseless dolphin slaughter. Other countries are not held responsible under United States Law, unless they are within U.S. waters.

With the pressure placed on the U.S. tuna fleet by the MMPA, many of the smaller companies sold-out their operations to foreign fleets. When American consumers found out about that, they began applying their own pressure to the tuna industry with organized protests, letter campaigns, and nation-wide boycotts at the grocers.

With little success in lowering dolphin mortality rates, Congress added more pressure to the tuna industry in three ways: 1) a ban on the seal-bombs thrown in the water to disorient the dolphins; 2) implementing an observer program, which actually placed national observers on tuna boats; and 3) threatening to implement legislation requiring "Dolphin-safe" labeling on all cans of tuna sold in the U.S. However, with increasing consumer pressure, the tuna industry finally voluntarily agreed to stop buying tuna from foreign fleets known to be associated with the killing of dolphins, and started labeling cans of tuna as "Dolphin-Safe."

After all that, there are still some twenty thousand dolphins dying at the hands of tuna fleets in the ETP. While the "zero" mortality rate is still pursued and projected. I have to say that twenty thousand dolphin deaths is better than one hundred thousand, but it is still not good enough. One dolphin senselessly murdered is one too many!

11 POLLUTING THEIR HOME

Pollution is simply the accumulation of any substance or material in a place where it isn't supposed to be. There is one obvious fact we must acknowledge, which is before humans began to multiply exponentially and spread across the face of the Earth there was no pollution. That is not to say humans are more powerful than God so as to have the ability to destroy the Earth. This is His favorite place in the universe, and He will not allow that to happen.

God created a perfect world! He wisely established all the elements along with their own regenerative cycles. He created all the animals and plants, endowed with the innate abilities of pro-creating kind after kind. He set up all the necessary food chains, intricate and inter-linked, which ultimately says God feeds the food that feeds us. He established the boundaries for the oceans. In His infinite wisdom and awesome power, He holds it all together and makes it work!

Far be it from me to question God's motives for He is incapable of error. It is impossible for us to understand God's entire plan and reason for allowing some things to happen. His mind is unsearchable! Nevertheless, He saw fit to bless us with this thing He calls, "free will." And in this free will, we can actually alter God's plan; that is, to the point at which He has had enough and puts His foot down with a vengeance.

Yes, God created water by combining hydrogen and oxygen, but He allows us to add to those elements anything we choose to; things which are soluble in water and things which are not. Nonetheless, in spite of our free will, God's command over the Earth still prevails!

It is said that the water on the Earth today is the very same original water that has always been here and the only water we will ever have. This gives more credence to sayings like, "If you dump it—you drink it," referring to used motor oil or toxic chemicals. At least we have some means of cleaning up the water we drink and use, but what about the aquatic life? They have to live day to day in what ever we humans dump into the creeks, rivers, lakes, and oceans and that is usually the exact progression of contamination.

The constant flowing, or movement, of water around the planet is called the hydrologic cycle. But to gain a better understanding of that cycle, perhaps a minor lesson in the properties of water is needed first. Because dolphins are conceived, born, and live their entire life in water and their food comes from the water, understanding the properties of water will help you better understand water pollution as well as water's unique role in living things:

1) water molecules are highly cohesive (sticky) due to their polar (+,-) nature;

- surface tension creates a membrane on the surface, which allows some organisms to actually walk on water.
- b) water siphons up tree trunks.

2) water is known as "the universal solvent."

3) water has a high heat capacity and heat of evaporation.

4) ice is less dense than liquid water, which is why it floats rather than sinks.

Each water molecule is constructed with three atoms: two hydrogen atoms and one oxygen atom. The oxygen atom has two rings where electrons orbit the nucleus; two electrons occupy the ring nearest the nucleus and six electrons occupy the outer ring, which gives the oxygen atom a negative (-) charge. Each of the two hydrogen atoms have one electron ring with a single electron orbiting each nucleus, which gives each hydrogen atom a positive (+) charge.

The two positive charged hydrogen atoms are attracted to the negative charge of the oxygen atom and each of the hydrogen atoms share electrons with the oxygen atom forming covalent bonds. The water molecule takes on a "V" shape with the hydrogen atoms clinging to the oxygen atom at a 105 degree angle. The unique "V" shape of the water molecule is what gives it a polar nature and makes each molecule ready and available to "stick" to other molecules through hydrogen bonds. Through hydrogen bonds (one-tenth as strong as covalent bonds), each water molecule can bond with four other water molecules. Let me reiterate that a molecule of water is formed by covalent bonds, but multiple molecules are held together by hydrogen bonds. See Figure 8 (Postlethwait 1989, p.34).

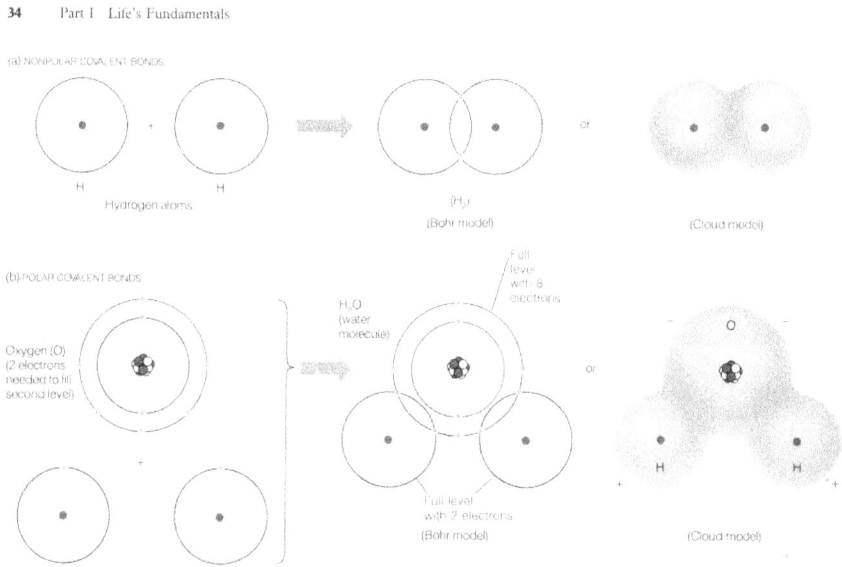

Figure 2.8 Covalent Bonds. (a) As hydrogen atoms approach each other, their individual electron orbitals overlap, fusing into a single molecular orbital. Because the distribution of charge is symmetrical, the bond is called a nonpolar covalent bond, and the molecule is electrically neutral. (b) In the H_2O molecule, the oxygen nucleus with its six protons attracts the two hydrogen electrons more strongly than do the hydrogen nuclei. As a result, the shared electrons spend more time orbiting the oxygen atom than the two hydrogens. This causes the oxygen atom to act as if it is slightly negatively charged (or, in chemical terms, **electronegative**) and the two hydrogen atoms to act as if they are slightly positively charged (**electropositive**). This kind of bonding is called a polar covalent bond, because regions (poles) of the molecule have slight positive or negative charges.

Figure 8

Without getting into the mathematics of energy amounts required to change the states of water from solid to liquid to vapor let me say this; when water freezes, air is trapped between water molecules making ice less dense than liquid water. Therefore, ice floats on the surface; and ponds, lakes, and seas freeze from the top—down. This is a good thing because aquatic organisms can carry on with their lives under the ice. God knew what He was doing. It is no random accident in nature.

The polarity, due to the hydrogen bonds of water molecules also allows for water to maintain a high heat capacity. That is, water has a way of holding accumulated heat, which keeps our oceans from boiling away or freezing solid. The heat capacity of water allows the oceans to help maintain the Earth's temperature throughout the various seasons. For example, an African dessert can reach 50 degrees centigrade while at the same time in Antarctica it can be -50 degrees centigrade giving a range of 100 degrees. However, oceanic temperatures vary from an almost constant 28 degrees centigrade at the equator to -2 degrees centigrade in Antarctic waters for a range of only 30 degrees.

The polarity of water molecules also provides for its excellent dissolving abilities. The "V" shape of a water molecule allows it to form hydrogen bonds with molecules having a different chemical make-up other than that of water.

The movement of water through the Earth's reservoirs is called the hydrologic cycle. That is, around the world water is constantly changing states from solid to liquid to vapor, and when it does it changes location. Somewhere around the world water is evaporating from a liquid reservoir. As it rises through the atmosphere it condenses and falls back to the Earth as either rain, snow ,or sleet.

When it falls on land it runs-off into a stream, which leads to a river, which may run through a lake, and eventually returns to an ocean, simply put.

Water covers seventy-one percent of the Earth's surface. The places where water resides, in whatever chemical state it is in, is called reservoirs.

Duxbury and Duxbury (1989) provide the following information regarding the hydrologic cycle: "About 396,000 km^3 of water move through the atmosphere each year. 334,000 km^3 are evaporated from the oceans while the land loses 62,000 km^3. When the water returns as precipitation, 297,000 km^3 are returned directly to the sea surface, and 99,000 km^3 return to the land. However, the excess gained by the land (37,000 km^3) flows back to the oceans in the rivers, streams, and ground waters of the world."

You can see that the hydrologic cycle is a dynamic one moving large amounts of water around the whole Earth annually. It is the Earth's natural way of cleaning itself. However, with human populations growing exponentially, pollution is concentrating in the bays, estuaries, and oceans of the world. For example, near the end of winter, early spring, the majority of households dump pounds and pounds of fertilizers (including phosphates and nitrogen compounds) on their lawns. Then a heavy rain comes and washes all that fertilizer into the run-off sewer system, which runs into a nearby stream, which runs into a nearby river, which flows into a bay or gulf, which is connected to an ocean. It should be easy to understand that lawn fertilizers not only work well for lawns, but also work well for algae, moss, phytoplankton, and other aquatic flora throughout the water systems of the world. See Figure 9 (Duxbury 1989, p.18).

Reservoir	**Volume (km3)**	**% of Total volume**	**Sphere Depth (m)**
atmospheric water	15.3 x 10^3	0.001	0.03
rivers and lakes	510.0 x 10^3	0.036	1.00
ground waters	5,100.0 x 10^3	0.365	10.00
Glacial and other land ice	22,950.0 x 10^3	1.641	45.00
oceanic water and sea ice	1,370,323.0 x 10^3	97.957	2686.00
Totals	1,398,898.3 x 10^3	100.000	2742.03

Figure 9

Without involving toxic chemicals, ponds, rivers, lakes, bays, and estuaries (where some dolphins live) can become killers to the animal life within through a process known as eutrophication (nutrient-rich water). While eutrophication is a naturally slow process in the life-span of water bodies, usually due to the build-up of silt and sediments, it is more profound after dramatic increases of nutrients, which cause algae and phytoplankton populations to explode.

Algae and phytoplankton are either suspended in the water column or floating on the surface. The explosion of their growth blocks sunlight from benthic plants (bottom-rooted plants). Or visa-versa, if the nutrients are located in the sediments, benthic plant growth explodes and choke-out the water column. Either way, the result of eutrophication is a disruption in levels of dissolved oxygen (D.O.). Dissolved oxygen is oxygen in the water that is ready and available for use by the biological organisms in the water. D.O. levels are needed in a body of water to sustain life.

The specific amounts of D.O. needed by organisms to survive in a body of water is called the Biological Oxygen Demand (B.O.D.). When D.O. levels fail to meet B.O.D. expectations, fish and other animals begin to die. Eutrophic lakes are termed "dead." If bays and estuaries supporting populations of dolphins become eutrophic, the dolphins are forced to move elsewhere to find food. Dolphins do not depend on D.O. directly because they breathe air the same way you and I do, but their food source is directly affected.

Now let's add the element of toxic chemicals to the hydrologic cycle. In 1961, Austin, Texas was a quiet little city. There was no industry to speak of, only the state government and the University of Texas. Austin was a blue-collar town supported by restaurants, entertainment, and construction workers. But on January 15, the residents of Austin woke up to the smell of dead fish. Tons of dead fish, turtles, crawfish, and snakes covered the water's surface of Town Lake and the Colorado River for about five miles downstream.

The ensuing investigation led agents of the Texas Game and Fish Commission to the plant where the plant manager admitted discharging the chemicals into the storm sewer and that the procedure had been practiced for ten years before.

"Game Warden Grover Simpson said Tuesday that Acock Laboratory, 2700 East Fifth Street, paid a $115.50 fine in the Justice Court of J.H. Watson on a charge of washing fish-killing chemicals [chlorinated hydrocarbons - Chlordane] into storm sewers which lead into the Colorado" (The Austin Statesman, 1961).

By January 21, the blanket of death had traveled through one hundred miles of the Colorado River and continued its journey towards the Gulf of Mexico. The gulf coast region at the mouth of the Colorado River is a complex ecosystem where fresh water begins to mix with salt water. The Intercoastal Waterway crosses the river at Matagorda, TX about a mile before its mouth to the gulf. To the north of that intersection is Matagorda Bay with its oyster beds, shrimp fisheries, and a significant sport fishery for the state of Texas. As the blanket of death approached the coast, officials decided to shut-down the Intercoastal Waterway by closing the lock system and allow the river of death to flow directly into the gulf.

After the cleanup of dead fish, Fish and Game technicians conducted a trawl of the river to see if any life remained, their nets were empty (Carson, 1962).

Following the catastrophe, the Texas Health department issued "Warnings" that any fish caught in Town Lake at Austin should not be eaten. That warning remained in effect until the year 2000.

The hydrologic cycle not only carries our chemical pollution to dolphin waters, but also our trash and debris. That soda can, cigarette butt, or piece of plastic you threw out the window of your car will end up in the gutter. When a hard rain comes, it will be washed into a creek, then a river, then a bay or estuary, then the ocean.

My favorite college biology teacher, Dr. Yvonne Estes, reminded us often that, "there is no such place as away." When you think you are throwing something "away" you are not; you have simply moved it from where it was to where it is. Have you visited your local landfill, lately? It is full of the stuff you threw away. It hasn't gone anywhere, it is still right there and will be for a long, long time.

Of course, plastics are the worst because they do not break down in the environment by natural process. Do you want to know how that affects dolphins and whales? Write, or call, your nearest Marine Stranding Network and ask them for a list of stomach contents of one or more dead Cetaceans washed up on the beach; you may be surprised to find that, that list will probably coincide, at least partly, with the contents of your kitchen garbage can. The amount of plastic used to package the things you and I buy in our daily lives is increasing dramatically from year to year. "In 1975 nearly 5.6 billion pounds of plastics were used in packaging, in 1987 this figure increased to 15.2 billion pounds" (O'Hara, et.al., 1988).

Plastic six-pack rings are very bad for young dolphins (not to mention the rest of the critters who live in the sea), who like to play with almost everything they see in the water. They use their rostrum to feel and to move things around in the water. Can you imagine the problem they encounter when their rostrum gets stuck in one of the holes of a six-pack ring? It could be the death of that young dolphin. "Along 300 miles of Texas coastline, more than 15,600 six-pack rings were found in three hours" (O'Hara, et.al, 1988).

For larger whales, the problem is plastic bags and sheeting. "During a 150 mile survey of North Carolina beaches, more than 8,000 plastic bags were found in three hours" (O'Hara, et.al., 1988). Sometimes the plastic is ingested, mistaken for jellyfish. "One turtle was found with 15 bags in its stomach, a whale was found with 50" (1988). Next time you see millions of balloons released into the air at an event like the Superbowl or the Indy 500, ask yourself how many Cetaceans and other sea life will die by choking on a piece of rubber from such an act. What goes up must come down. Sure the balloons are pretty as they disappear into the bright, blue sky, but upper-level wind currents can carry them hundreds of miles before they fall back to the Earth. And where do you think they are going to land? That's right, in the ocean.

Let's not forget the huge amounts of garbage (paper, glass, rubber, metal, and plastics) dumped directly into the oceans by humans. Some of these sources include, but are not limited to, merchant marine shipping fleets, commercial fishing fleets, passenger cruise lines, military vessels, recreational boaters, oil drilling rigs, and vessel accidents occurring at sea. In 1975, "the National Academy of Sciences estimated that ocean sources dumped 14 billion pounds of garbage into the sea every year—more than 1.5 million pounds per hour" (O'Hara, et.al., 1988):

SOURCE POUNDS

Merchant Shipping

Crew Wastes..............................242,550,000

Cargo Wastes........................12,348,000,000

Commercial Fishing

Crew Wastes..............................749,000,000

Fishing Gear..................................2,205,000

Recreational Boating.................227,115,000

Military.......................................163,170,000

Passenger Vessels.......................61,740,000

Drilling Platforms.........................8,820,000

Other..220,500,000

Total......................................14,023,800,000

Noise pollution is another consideration regarding dolphins and whales. As I said before, the U.S. Navy started experimenting with dolphins because the audio listening systems used by the Navy was unable to distinguish between motor sounds and whale sounds. Research has shown that blue whales have the capability to communicate over some two thousand miles with each other by emitting low frequency sounds at great depths, which travel with currents in certain layers of the oceans such as thermoclines.

Surely you must realize we humans are a noisy species. This is easy enough to determine. Leave your congested city or even your small town. Drive several miles out of town until the traffic begins to diminish. Find a small, two-lane country road leading out into the wilderness. When you get to a place where you see no other humans or houses stop your car and get out, listen and count the sounds you hear. You may hear a bird, a cow, a horse, or the wind. You may hear that overwhelming sound of silence.

The sea is naturally a noisy place, remembering that sound travels through sea water at an average velocity of "5,000 ft./sec. as compared with 1100 ft./sec. in dry air at 20° C" (Duxbury and Duxbury, 1989). The distance sound travels in sea water increases with temperature, salinity, and pressure, and decreases respectively. Low frequency sounds travel farther because they cause friction between water molecules as they absorb the energy, whereas high frequency sound passes through water vibrating the molecules.

In sea water, you have the sound of waves breaking, crabs and shrimps snapping their claws, the shifting sands and sediments, and sea birds crashing the surface for food. Add to that the sounds of human encroachment: people playing in the surf on beaches, fishermen dragging their lures through the water, boats, barges, ships, hammers and drills banging out new docks and piers, and the outrageous sounds of steel against steel, and steel against earth emitted by drilling rigs. What is a dolphin to do? I wonder if dolphins get headaches from all the noise?

Yes, God created the Earth and He alone holds it together, but, we are supposed to be stewards of the Earth. It is our responsibility to take care of it and respect it. God uses the Earth to feed us and provide us with the things we want and need throughout our lives.

Come on people, use your brains. Don't be a fool and waist or abuse the things God has provided and allowed you to borrow during your short time on the Earth. Take care of this Earth, it is the only one we have and you are not the only human on it!

12 CAPTIVITY

Of all the discussions concerning dolphins, perhaps none raise the emotions of humans as much as keeping dolphins in captivity. While you or I may form a personal opinion about the matter, there is no clear yes or no answer to the debate. Each side has its own good and bad points. And to those points, good or bad, there are exceptions to the rule.

Therefore, let me begin by saying, in general, I am against keeping dolphins and other marine mammals in captivity, but that in no way is a conclusive answer. I can agree with holding them captive solely for the pure purpose of legitimate scientific research, and as I said there are good and bad points about that. On the other hand, I can profoundly state that I am definitely and totally against any person or organization who publicly displays the animals for monetary profit. There are no good points to that argument. Now my task is to explain my reasoning.

There is no doubt that major contributions towards our knowledge of Cetaceans have come from scientific research through captive situations by noted scientists like Ken Norris, Ken Martin, Louis Herman, and Diana Reese to name a few. By keeping certain, few animals in captivity, scientists are able to gain information about dolphins that may, I say may, not be possible in the wild. Don't forget, there are scientist studying dolphins in their natural habitat—the open ocean and bays around the world.

When I began studying dolphins in 1989, I had no idea where my studies would lead me or even what it was I wanted to learn about them. As it was, or is, with any other subject, the more information I was exposed to, the number of questions in my mind increased.

Having no one first-hand to answer these questions directly, finding the answers I sought was like walking through a park on a moonless night without a flashlight; seeing a dim light in the distance through the trees, but not knowing what obstacles would hinder arriving at that destination. I knew I would eventually find the answers, but I didn't know where to begin searching. I can now say this with certainty, as knowledge of a particular subject increases, the distance between the next question and the answer decreases.

During my first year and half of researching Cetaceans, I became aware of the debate concerning their captivity. This opened a whole new field of research for me because the subject not only involves science, but also philosophy and ethics. The debate, by itself, is puzzling and very troubling to the mind, but what really saddened my heart is when I started seeing pictures of individual dolphins suffering in abusive situations with no apparent benefit to either the dolphin or the holder.

You and I know that when we receive broad, general information about something, which is supposed to be bad news, but really doesn't affect us directly, it is easy to let it go in one ear and out the other. Take for example famine; when we hear, perhaps on the car radio, of massive populations of people starving to death in a foreign country we think to ourselves, "Man, that's bad. Somebody ought to do something about that." Then you get home, turn on the television, and there is the same story, but this time with video of little children and babies, even adults with flies swarming their deformed bodies because they haven't had good food to eat their entire lives, and the story takes on a different meaning.

After reading research papers and periodicals where this person says "this" about keeping dolphins in captivity, and that person says "that" about the subject, I finally saw a documentary on the debate of captivity in which a solitary bottlenose dolphin was shown in a pool at a hotel (they didn't say where, understandably) with no companion other than a blown-up doll of a crocodile and a sign saying, "NO SWIMMING! Do Not Feed Or Touch The Dolphin!" The water was lime-green with algae. I have to tell you, that scene wrenched my heart and made me cry like a baby.

Later in the show, "Flipper's" original trainer appeared holding a list of dolphins held in captivity within U.S. jurisdiction and explained it was from The National Fisheries Service. That was my key to open another door to the debate over dolphin captivity.

With a little more research, I found the address of The National Marine Fisheries Service (NMFS). So I fired off a letter with my request for a list of dolphins in captivity and within a few weeks it arrived.

The Marine Mammal Inventory Report (MMIR) came with a list of individuals or organizations with permits from the NMFS who actually have Cetaceans in captivity. There may be others who have permits, but do not actually have animals for that particular year. Furthermore, the list included places in foreign countries with Cetaceans. Regarding such places, their obligation to report captive Cetaceans in their care to the NMFS is limited to animals imported or exported to or from the United States.

The MMIR also contains a code sheet to explain the different categories of information found in the report. There are eleven different categories of information on each sheet of the MMIR: ANIMAL NAME/IDENTIFICATION, SEX, ESTIMATED BIRTH YEAR, AUTHOR DOCUMENT, DATE TAKEN OR ACQUIRED, TAKE TYPE, LOCATION OF TAKE (PLACE NAME AND LONGITUDE-LATITUDE), COLLECTOR OR SOURCE, CURRENT STATUS, DEATH OR DISPOSITION (DATE/EXPLANATION), NECROPSY FILED W/NMFS. Basically speaking, any Cetacean captured in U.S. waters, held in captivity in the U.S., imported into or exported out of the U.S., born in captivity in the U.S. or held in captivity after being rescued for any reason in the U.S. must be documented on a MMIR form for that particular year.

There is so much information to be gained from the MMIR that many hours of study are required in order to make having a copy useful. Furthermore, having just one year will provide information regarding a particular institution or animal for that year. However, after receiving the MMIR for several consecutive years, one can begin to put the pieces of the puzzle together; you can begin to see patterns used by certain institutions for acquiring animals and indications of how well they manage the animals' care, and you can track individual animals as they may be transferred from institution to institution or as their health may change from "good" to "poor."

On the other hand, our Constitution says the government is of the people, by the people, and for the people. Therefore, we the people constitute the government's watchdogs, and there is no other way to be a watchdog than to obtain the information needed to monitor the operation of our government. So as an American citizen you do have the right to obtain this type of information. It is my opinion that an uninformed, uneducated public allows its government to manipulate it without scrutiny or voiced preferences.

Furthermore, if you rely on the media as your sole source of information as to the actions taken by our government, you are basically uninformed and uneducated about the laws of our land.

I strongly suggest writing a request to your representative in Congress (see Appendix "A") for "A CITIZEN'S GUIDE ON USING THE FREEDOM OF INFORMATION ACT AND THE PRIVACY ACT OF 1974 TO REQUEST GOVERNMENT RECORDS." The title is self-explanatory. The guide also provides information concerning "fees" for obtaining information from government agencies.

In a letter I received from the Chief of the Permits Division of NMFS, I was told, "If the information is over 288 pages, you will be charged 0.7 cents per page for the copies" (Turbush 1994).

After extensive homework and compilation, I can tell you that the 1992 MMIR reveals this, as well as much more, information:

Number of Institutions Holding Marine Mammals Captive:

- Public Display - 56 (of the 56, 24 are foreign)
- Scientific Research - 4 (of the 4, 1 is foreign)

Total Number of Animals Held Captive: 774

Number of Species Held Captive: 15

- **(Common Name):**
- Beluga - 36
- False Killer Whale - 23

- Killer Whale - 26
- Short Finned Pilot Whale - 3
- Pacific White-Sided Dolphin - 29
- Common Dolphin - 5
- *Atlantic Bottlenose Dolphin - 533
- Atlantic Spotted Dolphin - 1
- Commerson's Dolphin - 9
- *Bottlenose Dolphin (sp.) - 90
- Risso's Dolphin - 3
- Guianna Dolphin - 2
- Boutu Porpoise - 3
- Indo Pacific Hump-Backed Dolphin - 1
- Irrawaddy Dolphin - 10

You can see by the asterisks that bottlenose dolphins by far out number all Cetacean species held in captivity.

Remember what I said about longevity of study being important. After receiving several years of the MMIR, you begin to see patterns and discrepancies. It wasn't until after several years of receiving the MMIR that I learned I had the opportunity to request certain information in a MMIR. For example, the 1992 MMIR provided no information on animals that had died in captivity and the list of holders distinguished between "Public Display" facilities and "Scientific Research" facilities. The 1993 MMIR also provided no information on dead animals, but the distinction of holders was only those "with dolphins and/or killer whales," which excluded all other Cetaceans like False Killer Whales, Beluga Whales, and porpoises, etc. The 1994 and 1995 MMIR's provided no distinctions between holders, but did provide information on dead animals and necropsies filed with NMFS.

The point to note here is if you choose to request a copy of the MMIR be sure to specify the particular information you desire (e.g., year, species, information separated by species or by holder, etc.). Make your requests as specific and detailed as you can and be creative.

Since 1995, following an amendment to the MMPA of 1972 in 1994, there have been some changes at NMFS regarding permits and the MMIR. For one thing, "PERMITS" are no longer issued by NMFS for publicly displaying Cetaceans. Instead, the United States Department of Agriculture (USDA) issues a license to a "Holder" or "Facility" under the Animal Welfare Act through the office of Animal and Plant Health Inspection Service (APHIS). Furthermore, APHIS now controls the care of all captive animals, be they rats, monkeys, or Cetaceans.

The license issued by USDA-APHIS for the purpose of public display is comprised of three criteria: The animals must be under the care of a veterinarian (graduated from an accredited school of veterinary medicine), must provide an educational program, and must be open to the public on a regular basis. Also, NMFS received new software changing the MMIR format.

"AMBASSADORS" FOR THEIR KIND

Some people go through their entire life never noticing wild animals right in front of their faces. Many could care less about the wild animals that share the planet with them. Others would like to see wild animals, but are never in the right place at the right time. Sometimes, human eyes are open, but they do not see.

Large amounts of money are spent by people annually to insure a wild animal experience. Of course, zoos probably have the widest variety of species for viewing if you just want to see what a particular animal looks like, but not learn anything about its true behaviors in its natural habitat. I would venture to say the stripes on a zebra look very much the same whether you are in San Antonio, Texas or the grasslands of Africa.

I also have to go a step further, though, and say that the behaviors of a zebra in the San Antonio Zoo differs greatly from that of a zebra in Africa, duh. I must even go a step further and say that the zebra stripes in the zoo may begin to change once human beings start playing with genetic engineering, which is inevitable when you add "Ph.D." suffixes to names of humans with no morals when management turns into manipulation.

It is easy to understand how we came to live in a society where some people still think beef and chickens are produced in a factory and shipped to the grocery store. People drive down the highway and see pasture after pasture full of cattle and tell their kids, "Look at the cows . . . remember that fairytale book I read to you . . . well there's the cows." While the kids are snug in their beds late at night, they never see the eighteen-wheelers rolling down the road carrying those "cows" to the slaughterhouse. They never peer through the panels of the trailers and see the youngest, or the weakest, animal collapsed on the floor being trampled by the bigger ones, or the one whose eye has been gouged out by the horns of another. That is not a pretty picture to see in your mind while you're eating a hamburger.

Our society is raising generation after generation of people who will never step foot inside a slaughterhouse. Unless someone tells them at some point in their lives that those "cows" they saw in the book or out in the field will some day be led down the shoot, harnessed, then struck in the head between the eyes with a sledge-hammer or shot with a small caliber gun so as to stun, but not to kill, then have its throat sliced open so the animal will bleed to death, most kids will grow up thinking steaks are produced in a factory rather than in the womb of a female bovine.

Don't get me wrong, I eat meat, but I know where it came from and that God gave us certain animals to eat (Leviticus 11).

Therefore, we can look at "cows" in two different ways: as an animal living its life in a field, or as a steak on the grocer's shelf. The question I pose to you is from which viewpoint would you consider the "cow" to be an "ambassador for its kind?" Does the cow in the field tell the whole story about a cow's life or does one learn about a cow's life by eating a steak?

A simple definition of "ambassador" is: an authorized messenger or representative. The word "authorized" connotes choice, privilege, and appointment. Therefore, "ambassador" is a position sought after and accepted. It is a position held in high esteem, one that takes a message of hope and positive thoughts from whence it came to where it now resides. It is a position that conveys a way of life and proclaims a story of lifestyles of those being represented.

To the best of my recollection, the phrase "ambassador for their kind" had not been used with regards to animals until it was applied to Cetaceans being held in captivity. The entertainment industry, exploiting animals, needed a catch-phrase to counteract allegations against their actions. In other words, oceanariums and aquariums were turning dolphins into circus animals and they were met with strong opposition, which said "what you are doing is wrong." So the oceanariums and aquariums came up with the idea to call the dolphins "ambassadors for their kind."

Here is the trick that fools the public, one of the pre-requisite orders of the MMPA for publicly displaying Cetaceans is: Sec.104, (©)(2): "A permit may be issued for public display purposes only to an applicant which offers a program for education . . ."

The paragraph does not say how much education, what type of education, or what the content of the education information should be. What does an ambassador do? It educates foreign governments of the policies held by the land from whence it came.

I must admit I went to Sea World of Texas once to see what they do there, and yes, they did educate me—they told me a dolphin is not a fish, it is a mammal. They also told me when a dolphin jumps out of the water it is not to be called "jumping," it is to be called "breaching." Wow! That is some revolutionary educational information.

After one of the shows, a fellow spectator requested the company of one of the trainers as they made their way from the arena. I recognized the opportunity to interact with someone who should have first-hand, up to date information concerning the animals he commanded. So I went over to participate in the impromptu "Q & A" session.

"Do you ever have animals that don't want to perform?" I asked.

"Well . . . you can't put that kind of anthropomorphism on these guys. We interact with these animals every day, and if we think an animal is acting strangely we get the vet to check them out."

"Are any of these animals related?"

"Only if they were born here, then after a period of time we exchange them with animals from another facility to prevent in-breeding," he said.

"You mean even though delphinids are social animals and naturally exist in extended families, your policy is to break-up such relationships?"

"Well, I guess you know this is not their natural habitat. These holding tanks are confined spaces and we can't control all of the animals' natural instincts. These are wild animals."

"Where do you get your animals?" I asked.

"They come here by legal means . . . I'm sorry, I have to go."

Now you have an example of the kind of information you are likely to receive when you visit a dolphin show; while they are willing to give you a "second grade" education about dolphins because they are required to by law, they are un-willing to divulge information, which would be detrimental to their multi-million dollar industry.

They tell you that a dolphin is a mammal, but they do not tell you delphinids, including killer whales, exist naturally in extended families and that in the wild their social groups have their own ways of dealing with in-breeding, which does not decimate family ties.

They tell you the animals are fed adequately whether or not they perform, but they do not tell you that delphinids eat live prey in the wild, not dead, frozen fish. If they tell you Cetaceans are predators, then they will say they have actually done the animals a favor by eliminating the need to search for food. How devious?

They tell you the animals receive the best of veterinary care, but they don't tell you that wild dolphins live thirty, forty, or fifty years in the wild without ever having to go to a doctor.

They may tell you about their wonderful breeding program, but they won't tell you how many animals were traumatized or killed in the process of capture to obtain the breeding-pairs.

They tell you Cetaceans are sonic creatures, but they don't tell you that placing such a creature in a concrete impoundment is an abusive form of sensory deprivation. You will never hear a representative of Sea World say, "*Killer whales will most likely stay with their relative family their entire life. Some are members of a pod whose range is one or two hundred miles radius. Others may travel thousands of miles. And others may travel the entire globe, but we've stolen this particular one from its family and confined it here in this relatively small swimming pool, and we appreciate your paying $35 per person to come watch us play with it.*"

The people who convert wild dolphins into "circus clowns" refuse to tell you the whole truth about the animals they imprison because doing so would jeopardize their hundred million dollar per year industry. Like a politician, their answers are vague and diverting

Of course, it would be wrong to place all the blame on a company like Sea World Inc. After all, we all have to work somewhere to earn a living. All they are doing is filling the need of the public—the need to be entertained. They just do it better than anyone else.

Face it, people do not go to a place like Sea World to be educated about Cetaceans. They go there to be entertained by them! By and by, we humans are a lazy sort. We could go to any library, check out any children's book on Cetaceans, and learn more about them than we will at Sea World. Or we could drive a few more miles and watch wild dolphins playing in their natural habitat at no charge. But that would require too much thought and planning and work. So instead we load up the car with our kids and our wallet and take them to a place that provides us with the entertainment we need. That's much easier, no matter the cost.

A brief visit to reportgallery.com\anheuser-busch\annual reports\entertainment section (Anheuser-Busch is the parent company of Sea World Inc.) reveals that in 1997, 21 million customers visited the nine entertainment parks yielding over $100 million in profits. In 1998, 20 million customers and $116 million in profits. In 1999, 19 million customers and over $100 million in profits. You can see that the American public has supplied a substantial market for being entertained by imprisoned and enslaved Cetaceans. I also queried five other companies regarding their profits, but all five declined to share that information.

Companies, such as Sea World, claim their captive dolphins and killer whales are "ambassadors for their kind." Let me ask you this question: are you willing to be kidnapped away from your family and friends to spend the rest of your life incarcerated in a Russian prison disguised as an "ambassador for your kind?" Maybe you react to that question by saying something like, "it can't be all that bad for those dolphins and whales." Well, have you spent any amount of time incarcerated, unable to leave when you wanted to, eating only what you are given to eat, and isolated from other humans you know and love?

It has been established that Cetaceans are intelligent, even sentient beings. Do you think they have no comprehension of the brevity of incarceration? Don't give me that "anthropomorphism" crap that proponents of Cetacean captivity use as an excuse: *they're just animals, they don't know what's happening to them; they can't sense being away from their pods or schools; they can adapt to life in a swimming pool; they don't care what they eat as long as they have a full belly; they do not have the ability to distinguish sadness, loneliness, or abuse.* Poppy-cock! I am not speaking here of dolphins held in captivity for legitimate research by scientists like Ken Norris, Ken Martin, Louis Herman, or Diana Reese. I'm talking about dolphins held in captivity for the purpose of "public display".

However, it becomes increasingly true that humans, certainly possessing the ability of gaining intelligence, have the ability to adapt to the same consequences suffered by Cetaceans held in captivity.

That is, when humans are involved in a project, if they don't like what they see they make changes to gain the proper results. The proponents of "captivity" would have you believe there are no detrimental affects on Cetaceans in captivity; I tell you they are wrong and they do not tell you the truth. For example, let's follow the captivity of two famous killer whales, Orky and Corky.

ORKY AND CORKY

Both Orky (male, NMFS #SWC-OO-8726) and Corky (female, NMFS #SWC-OO-8727) were captured "pre-act" (P/A), before the MMPA of 1972. Therefore, throughout their captivity there is no record available (N/A) of who captured them or where they were captured, and they were never counted against any quota sanctioned by NMFS under the MMPA. In other words, a "permit" was never issued for either their capture or their confinement in captivity.

We do know Orky (SWC-OO-8726) was captured 5/10/68. Corky (SWC-OO-8727) was captured 12/27/69. Orky and Corky began their captive life at Twentieth Century-Fox Marineland Inc. in Rancho Palos Verdes, California. On 7 November, 1977 the facility became Hanna-Barbera's Marineland. While the two whales remained at the same facility, I doubt a change in command in the front office could be smooth enough to not effect at least one aspect of the animal's care, i.e. change in staff, change in procedures, or change in products.

In 1977, Orky and Corky were the first pair of killer whales to give birth to a baby in captivity. However, the baby had brain damage and died several days after birth.

On 11 January, 1982 Hanna-Barbera's Marineland became Marineland Amusements Corporation. While still at the same facility, Orky and Corky gave birth to four more babies; the last one in 1986, none of the calves lived.

The staff at Marineland considered the five failed births: 1) perhaps Corky had been stolen from her pod too early and never learned how to nurse a calf; 2) with only one tank, mother and calf could not be isolated; 3) the small size of the tank prohibited the natural feeding process (even hand feeding did not work).

In another incident, a whale was killed in a fight, which was attributed to a small tank creating an "un-social environment."

On 21 January , 1987 Orky and Corky became the property of Sea World of California. At the transfer of care and transport of the animals, Orky and Corky received their new identification numbers. Each holder has its own formula for identification numbers; here is the formula for Sea World:

The first three letters represent the facility:

- SWT = Sea World of Texas
- SWF = Florida
- SWO = Ohio
- SWC = California

The next two letters represent the genus and species:

- TT = *Tursiops truncatus*
- GM = *Globicephala macrorynchus*
- OO = *Orcinus orca*

The first two numbers represent the year care was initiated at that facility:

- 85 = 1985
- 86 = 1986
- 87 = 1987

The last two numbers represents the number of the animal, in numerical order, for care initiated to animals of that species for that year at that facility. Therefore, SWC-OO-8726 was the 26th killer whale whose care was taken over at Sea World of California in the year of 1987. Conceivably, there could also be an SWF-OO-8726 or an SWC-TT-8726, etc.

Eighteen months later (09/26/88), Orky died. As of 02/20/01, Corky (SWC-OO-8727) was still alive in San Diego.

In 1984, Sea World of Florida built larger tanks to accommodate up to four whales with secondary tanks to isolate animals. In 1985, they had the first successful birth of a killer whale in captivity (SWF-

OO-8501, 09/26/85). So my point is that animals (Cetaceans) may not always be able to adapt to the situations we humans place them in, but we humans have the ability to adapt to the situations required by the animals. But it took a long time for some human to finally figure out that there must be something wrong with placing a very large animal in a tiny swimming pool, and then actually make the changes. In that amount of time, how much did the animals suffer?

Still, there are those who say there is nothing wrong with incarcerating wild animals; remember that these are people in corporations whose annual profits continue to exceed one hundred million dollars. So is it true just because they say so? I think not! Let's first consider three aspects of captivity: 1) the type of animal; 2) captive-born animals; and 3) captured animals. Then we'll look at the rigors of captivity itself.

It can be said that a dolphin swimming around in the ocean is just another wild animal; while that is a true statement, it is not complete. I refuse to say that a wild dolphin is a "better" animal than a wild butterfly, but I can certainly say that a dolphin is much more biologically advanced physically and mentally than a butterfly. There are three particular conditions that support me on this: 1) the gestation period for dolphins is typically one year, that's three months longer than humans'; 2) parental care in dolphin families can range from two to five years where a baby must learn how to live its life as a dolphin; and 3) the social environment of dolphin societies requires that juveniles continue to learn from other dolphins (superiors and peers) the do's and don'ts of social hierarchy.

As for animals born in captivity I ask this question: Is it possible for me to be an "ambassador" for the United States of America if I have never, ever set foot in the U.S.? I think not! What we have now is a generation of captive born animals who have never tasted life outside of a concrete swimming pool. They are not "ambassadors for their kind." They are wild animals held in prison by humans. They are nothing other than circus clowns.

CAPTURING WILD CETACEANS

Now let's examine the capturing of wild animals. Before humans learned how to successfully breed Cetaceans in captivity, they had to actually go out into the open waters and physically capture wild animals. Before the MMPA of 1972, there was no permit required for this activity. However, while there have been no "capture" permits issued in recent years, I can give you several reasons why. First, the conditions of the MMPA (permits, reporting, and accountability) revealed the morbid truth about the harsh methods of capturing wild Cetaceans and the results of captivity. Let me support this by showing some real examples.

On Sept. 27, 1974 Sea World of California was issued a permit (#50) at the cost of $200 for the purpose of capturing six bottlenose dolphins and four pacific pilot whales for public display (see Figure10 on page 161). After looking at this real permit from NMFS, then look at Figure 11 (page 162) to find the results of that permit concerning the four pilot whales (you can refer to Figure 12 (page 163) and 13 (page 164) as a key to the codes).

U.S. DEPARTMENT OF COMMERCE
NATIONAL OCEANIC AND ATMOSPHERIC ADMINISTRATION
National Marine Fisheries Service

Permit for Taking Marine Mammals Permit No. 50

Sea World, Incorporated, 1720 South Shores Road, Mission Bay, San Diego, California 92109, is hereby authorized to take the marine mammals specified below for the purpose of public display, subject to the provisions of the Marine Mammal Protection Act of 1972 (16 U.S.C. 1361-1407); the Regulations Governing the Taking and Importing of Marine Mammals; and the special and general conditions hereinafter set out.

A. Number and kind of animals

1. Six (6) bottlenosed dolphins, either *Tursiops truncatus* or *Tursiops gilli*, may be taken.

2. Four (4) pilot whales (*Globicephala macrorhyncha* c.f. *scammoni*) may be taken.

B. Special Conditions

1. The dolphins, if *Tursiops truncatus*, shall be taken from offshore waters of the State of Florida and other southeastern coastal States, by means as stated in the Application. The dates and specific locations of such taking, and the desirability of National Marine Fisheries Service observers, shall be determined by the Regional Director, National Marine Fisheries Service, Southeast Region, Duval Building, 9450 Gandy Boulevard, St. Petersburg, Florida 33702.

2. The dolphins, if *Tursiops gilli*, shall be taken from coastal waters of the State of California and/or from coastal waters of Baja California, Mexico, by means as stated in the Application. If taken in territorial waters of Mexico, such dolphins may be imported.

3. The pilot whales shall be taken from coastal waters of the State of California, by means as stated in the Application.

4. The dates and specific locations of the taking of *Tursiops gilli* and *Globicephala macrorhyncha*, and the desirability of National

Figure 10

MARINE MAMMAL INVENTORY REPORT
Date of Report: 03/08/94

Page: 7

NAME OF ANIMAL HOLDER: SEA WORLD INC

SPECIES SCIENTIFIC NAME: GLOBICEPHALA MACRORHYNCHUS
COMMON NAME: SHORT-FINNED PILOT WHALE (code=041)

ASN: 1	LEX: 3
ANREP: YES	FNUM: P2A

ANIMAL NAME / IDENTIFICATION	SEX	EST BIRTH YEAR	AUTHOR DOCUMENT	DATE TAKEN OR ACQUIRED	TAKE TYPE	LOCATION OF TAKE PLACE NAME AND LATITUDE-LONGITUDE	COLLECTOR OR SOURCE	CURR STAT	DEATH OR DISPOSITION DATE	DEATH OR DISPOSITION EXPLANATION	NECRP FILED NMFS
SWC-GM-8001	M		#252	01/08/80	HP	CA, 10 MI NW PT LOMA	SEA WORLD	D-C	05/15/80	RESPIRATORY ABSCESSES	YES
SWC-GM-7402	F		#50	10/31/74	HP	CA, CATALINA ISLAND	SEA WORLD	D-N	11/15/74	ENDOMETRITIS < mon	YES
SWC-GM-7602	M		#50	12/18/76	HP	CA, PALOS VERDES	SEA WORLD	D-C	09/25/78	UREMIA < 2 YRS.	YES
SWC-GM-7603	M		#50	12/20/76	HP	CA, PALOS VERDES	SEA WORLD	D-C	05/20/78	ABDOMINAL ABSCESS < 2 YRS.	YES
SWC-GM-7401	F		#50	10/23/74	HP	CA, CATALINA ISLAND	SEA WORLD	D-C	10/05/81	CHRON KIDNEY DISEASE & SENILE CHANGES 6 YRS	YES
SWC-GM-7501	M		#50	11/14/75	HP	CA, PALOS VERDES	SEA WORLD	D-C	04/10/76	PNUEMONIA 5 mon.	NO
SWC-GM-7502	F		#50	11/16/75	HP	CA, PALOS VERDES	SEA WORLD	D-N	12/10/75	< mon.	NO
SWC-GM-7503 SWC	M		#50	11/21/75	HP	CA, PALOS VERDES	SEA WORLD	D-N	12/23/75	PNUEMONIA 1 YR.	NO
PW-72	F		P/A	11/15/72	HP			T-N	07/30/73	SOLD TO MARINEWORLD	N/A
SWC-GM-8002	F		#252	12/11/80	HP	CA, 12 MI NW PT LOMA	SEA WORLD	D-C	10/18/92	ACUTE NECROTIZING PNUEMONIA	YES
SWC-GM-8003	F		#252	12/16/80	HP	CA, 12 MI NW PT LOMA	SEA WORLD	D-C	07/05/83	METRITIS, PARASITISUM & PULMONARY EDEMA.	YES
SWC-GM-8227	M		#252	12/10/82	HP	CA, SANTA CATALINA ISLAND	SEA WORLD	D-C	12/19/82	CHRONIC CIRRHOTIC PNEUMONIA; PARASITISM	YES
SWC-GM-8226	F		#252	12/10/82	HP	CA, SANTA CATALINA ISLAND	SEA WORLD	G-C			
SWC-GM-8726	F		P/A	02/19/87	EX	CARE TRANSFERRED FROM MARINELAND AMUSEMENTS CORP (CAPTURED 9/6/66)	N/A	P-N			

Figure 11

I'm sorry, but you may need a magnifying glass.

In Figure 11, there are three columns of interest: AUTHOR DOCUMENT (#50); DATE TAKEN OR ACQUIRED; and DEATH OR DISPOSITION (DATE/EXPLANATION). In Permit #50, Sea World was allowed to capture four Pacific pilot whales, but they captured seven. Of the seven, two died in less than a month after capture; one died five months after capture; one died one year after capture; two died two years after capture; and one died six years after capture.

INSTRUCTIONS FOR

ANIMAL NAME/IDENTIFICATION
Provide the animal's name and ID number.

SEX M = male; F = female; U = unknown.

ESTIMATED BIRTH YEAR
To avoid having to update the animal's age every year, provide the estimated year of birth.

AUTHORIZATION DOCUMENT
Provide the Marine Mammal Protection Act (MMPA) authorizing document under which the animal is held. (Note: If captive born animal, use the same authorizing document as the female parent.)

- #____ = MMPA permit number; e.g. #52
- LE____ = MMPA Letter of Exemption number; e.g. LE42.
- AN____ = Agreement with NMFS; e.g. AN3C (If agreement number is not known, enter date, e.g. AN5/25/79).
- __/__/__ = Letter of authorization issued by NMFS, e.g. 7/25/78.
- State = Beached/stranded animal take by the State (or under authorization from the State) for rehabilitation purposes only and not for permanent maintenance.
- P/A = Pre-Act animal taken before enactment of the MMPA (December 1972).
- N/A = Take not covered by the MMPA; e.g. animal taken in foreign country and not imported to U.S.

DATE TAKEN OR ACQUIRED
Provide the date the animal was removed from the wild; or if obtained from another facility, provide the date the animal was acquired.

TAKE TYPE
Provide one of the following two-letter codes:

- HP = wild caught in the U.S., permanently held
- HT = wild caught in the U.S., released or escaped shortly after capture
- KW = killed in the wild (including accidental deaths)
- LM = live animal imported into the U.S.
- BS = beached/stranded animal
- EX = exchange or transfer from another facility
- CB = captive born
- KC = wild caught in U.S., intended for terminal scientific research
- FT = taken and maintained outside U.S. jurisdiction
- NA = none of the above

Figure 12

)MPLETING INVENTORY FORM

LOCATION OF TAKE
As specifically as possible, provide the geographic location of take or beaching; if an exchanged animal, provide the source facility; if captive born, provide the name of parents.

COLLECTOR
Provide the name of the collector of wild caught animals.

CURRENT STATUS
Combine one of the following codes on status of the animal with a code on whether the animal was counted against a quota:

Status
G = Animal alive in good health
P = Animal alive in poor health
D = Animal died
R = Animal released or escaped
T = Animal transferred to another facility

Count
N = not counted against authorized quota, e.g., pre-Act animals
C = counted against NMFS authorized quota

DEATH OR DISPOSITION
Provide the date of the animal's death, transfer, escape, of release. If the animal is dead, provide cause of death; if transferred, provide the name of the facility to which it was transferred; if escaped or released, provide the location of the escape or release.

NECROPSY FILED WITH NMFS
For NMFS use only.

Please note that necropsy reports should be on file for all animals taken under the authority of the Marine Mammal Protection Act.

NOTE: All dates must be entered into the computer as MONTH/DAY/YEAR. Therefore, if no month or day is provided, or a time range is specified, the mid-point is used, as follows:

1978 is entered as 6/1/78
March 1969 is entered as 3/15/69
June 6-10, 1980 is entered as 6/8/80

Public Reporting Burden
Public reporting burden for this collection of information is estimated to average 2 hours per response per year to complete this reporting form. This estimate includes time for reviewing the instructions, gathering and maintaining the data needed, and completing and reviewing the collection of information. Send comments regarding these burden estimates or any other aspect of this collection of information, including suggestions for reducing this burden, to the National Marine Fisheries Service, F/PR1, 1335 East-West Highway, Silver Spring, MD 20910, and to the Office of Information and Regulatory Affairs, Office of Management and Budget, Washington, D.C. 20503.

Figure 13

Also in Figure 11, you will find part of Permit #252, which is far more gross than #50. Issued 1/12/79, Permit #252 allowed Sea World of California to capture the following: six beluga whales, six pilot whales, twenty common dolphins and/or Pacific white-sided dolphins, fifteen spinner dolphins (and/or Northern right whale dolphins, spotted dolphins, and striped dolphins), and fifty bottlenose dolphins, with forty to be returned to the wild within ninety days.

I could write a short book solely on the subject of Permit #252, but what I want you to notice are the numbers of animals here. All of these numbers are disturbing to me, but my special interest is in the "fifty" bottlenose dolphins with forty to be returned to the wild. Needless to say, and you can obtain the MMIR and see for yourself, many of the "forty" never made it back to the wild; they died at the hands of the thief that stole them from their natural world!

Why would Sea World need to obtain fifty animals for ninety days, then release forty of them back to the wild? It is my speculation that they needed to increase their supply of "eggs and sperm" to maintain diversity within the gene-pool of animals already in their custody for future generations. If, indeed, this is true, it would suggest the beginning of genetic engineering, and who knows where that will lead? Is Sea World the new "god" of Cetaceans?

THE PRICE OF AMBASSADORSHIP

Let me ask you a very serious question: have you ever spent the night in jail? Those words don't even sound good, do they? But if you have, or even know someone who has, you understand how disruptive it can be to your peace of mind and daily routines in life. The affects of such an event are everlasting.

If you were an ambassador representing the United States, you would be fortunate and experience the very best of all things life has to offer. You would certainly enjoy a life of luxury. On the other hand, if you were an "ambassador" for all of dolphin-kind around the world your life would be drastically converted from freedom to slavery.

Regardless of whether or not dolphins are currently being captured from the wild for public display, the fact remains that the majority percentage of the total captive population of Cetaceans remains to be those who were captured. For them, life has become an endless schedule of routines manipulated by humans whose only desire is to get rich.

One day they were swimming around in the open ocean chasing and herding large schools of live fish. The next day they were confined to an area by concrete walls, swimming in chlorinated fake sea water being fed nothing but dead frozen fish. Strange humans are now talking to them, yelling at them, blowing whistles at them, jumping in the water with them, and touching them constantly.

For them, there are only two chances for reprieve from imprisonment: 1) stay healthy, but don't perform. A sick dolphin will be further manipulated by humans, but a healthy dolphin who does not perform is very expensive to maintain and does not make the owner rich, and 2) death.

Unfortunately for the dolphin who survives the initial probation period of captivity, the cost of retraining for reintroduction back to the wild is higher than that to maintain the animal for the rest of its life. It is, therefore, doomed to live the rest of its life in a concrete swimming pool.

WHAT'S WRONG WITH THIS PICTURE?

We need to wake up and face the facts, people. Even though God is in control of the creation or extinction of a species, humans can and do apply pressure to His control. Our duty as human-beings is to play the role of "stewards" for the Earth and all of its inhabitants. We fail at that duty when we falsely manipulate or abuse or try to assume the control of any species from the hands of God!

God equipped Cetaceans so they could successfully live their life freely in the oceans of the world. They are intelligent, sentient, and sonic creatures. It is simply wrong for us to condemn any of them to live their life in a concrete swimming pool for either our pleasure or our financial gain!

Since, therefore, dolphins do not naturally occur in concrete swimming pools, why do we expect them to survive in them? We have a situation where corporations are profiting over $100 million annually, yet they still refuse to adapt to the natural needs of the animals making them rich—that is exploitation!.

When you visit one of these large oceanarium parks, you can walk from one building to another containing large aquariums, which contain different fish, mammal, reptile, or bird species. The important thing to note as you view these displays is that the operators of the facility have tried to replicate the animal's natural habitat.

While the surface you peer through may be a sheet of plate-glass, the environment behind the glass (above and below the water line) will usually have rocks, soil, sand and vegetation; remember, smooth concrete is not a naturally occurring element in any environment anywhere in the world. Remember, also, none of these animals, excluding bats, are sonic animals.

Then you make your way to the stadium for the dolphin show and you find these sonic animals performing their show in a smooth concrete swimming pool. It doesn't take a rocket scientist to figure out that the institution's concept of how to display the various animals at the facility is completely backwards!

While I am glad the more fortunate animals get to enjoy a better representation of their natural habitat, why are the more intelligent, more advanced animals forced to live the remainder of their lives in sensory deprivation in the confines of smooth concrete? I have to go back to the analogy of going to jail; jail is not pleasant, there are no amenities of home. It is a form of punishment. This is the same type of punishment we apply to Cetaceans when we place them in a smooth concrete swimming pool.

For a wild dolphin or killer whale, ninety-nine percent of their life experiences are gained below the water's surface. There are things for them to do down there, that is where they live. Below the surface is where they are born, feed, grow, mate, play, and do all the other things that make a dolphin's life complete.

In a swimming pool, the only action a dolphin ever sees occurs above the water's surface. There is nothing for them to do beneath, nothing to look at, nothing to chase, and nothing to eat. They have no need to use their sonar because in the first place there is nothing down there, and secondly the sound will bounce back and hit them in the face like a sledgehammer. Furthermore, with their very good eyesight, they can see from one end of the pool to the other. Since everything happens for them at or above the surface, that is where they spend all of their time. This is why you see the dorsal fin of killer whales laying over to one side; without the buoyancy of water, gravity pulls it down.

The curators of these very rich corporations say it would be too expensive to build natural environments large enough for the Cetaceans, and water filtration is a larger problem. Yet they continue to enjoy profits over $100 million year after year.

"SWIM-WITH" PROGRAMS

Swim-with programs are another form of captivity that concerns me. With further deregulation these programs are springing up across these United States and around the world. The problem, as I see it, is two-fold; many of these facilities claim their dolphins are free to leave any time they wish, and the programs provide the opportunity to fulfill the fantasies of millions of people who want to experience "being with" a dolphin in the water.

Many of these facilities are located on the shoreline of a bay or gulf and constructed with physical access to actual sea water. The facilities may or may not have a low-profile fence surrounding a particular area for the dolphins. Regardless, the operators of these kind of facilities usually claim, "*...these are not captive dolphins. They can leave if they choose to. They are not forced to stay here.*"

The fact of the matter is they are, indeed, captive animals. To begin with, the MMPA of 1972 prohibits any human within the boundaries of the United States from coming into contact with or harassing any marine mammal. The only way you can get in the water with those dolphins is if that facility is licensed by USDA-APHIS to maintain and care for captive animals. Truly wild animals require no license, but you can't harass them either!

Moreover, while there are very few, rare occasions scattered through the history of time where a wild dolphin has befriended a human (as mentioned earlier), the encounters have usually been brief and initiated by the dolphin itself and usually involved a single, particular human.

Furthermore, the dolphins at these types of facilities are hand-fed and trained. Sure, they may have the ability to swim out through an opened gate to freedom, but who among us would rather go out to work hard for our food when a free meal is constantly offered at the same location? And what if they did leave the facility for good? What are the chances of another wild dolphin swimming in and replacing it as "resident captive dolphin" on their own free-will? And how long do you think that facility would stay in business if people came, spent $50 or $75 or more to swim with a dolphin, but there were no dolphins to swim with?

What about the simple act of a human getting in the water with a dolphin, what could be wrong with that? Let me ask you this question: when you go to a zoo do you get in the pin with a chimpanzee or a giraffe or bear or ride a zebra? Why not? Why should that be any different with a dolphin? We're talking about an animal (a dolphin) who used to be, or still is, wild. It can weigh anywhere from two hundred to six hundred pounds. It can move through the water much, much better than we can. It has about eighty-two sharp, conical teeth, with which it could rid your body of almost any appendage it wanted. With its rostrum or its tail flukes, it can break any bone in your body.

I'm talking about possibilities here. Okay, so you get in the water with a captive dolphin and it chooses not to cripple or kill you, is there any possible thing you could do to harm it willingly or unwillingly? Before you get in that water are you going to completely and properly sanitize your body from all germs and parasites you may carry? I doubt it. Why would you think that animal is completely immune to any and every illness we humans can suffer? If you think they are, you should examine the cause of death of the hundreds or thousands of Cetaceans that have died in captivity and you will have a change of mind. Furthermore, are you the only human in the water with the dolphin? No, you're in a group of humans. How many groups does that dolphin entertain each day? How many days/months/years has that dolphin been entertaining groups of humans?

I have to say some fantasies in life are better left unfulfilled. Swimming with captive dolphins is one example. Why are humans so selfish?

13 CULTURE

Culture is one of those aspects of life no one person can ever control. These are the traditions, beliefs, and actions that a society is built on. While the word "culture" usually pertains to geographic regions it is not limited by physical boundaries. Culture is subjective, created by learned behaviors and passed on from one generation to the next; wherein, lies the underscoring theme of this chapter—old habits are hard to break.

Once a cultural fad becomes practice and is handed down from one generation to another, pressures from the outside world to change the practice is quickly labeled "prejudice." The one side of the issue pointing the finger at the other says, "We don't do that. We don't like that. Why should they be allowed to do that!" The attitude of the one side receiving the criticism is usually that of, "This is our business. Mind your own business. Leave us alone!"

The degree of criticism is based on whether or not the practice in question is confined to a specific region or affects others as well, and pressure increases whenever the practice involves harming some form of life. For example, let's say there are two distinct cultures and each one clothes their bodies with the same style of garments. However, one of the two chooses to wear their garments in-side-out rather than in-side-in. You can easily see that such a practice causes no harm to life and has no ramifications concerning world peace.

On the other hand, AIDS is spreading around the world consuming those infected. One theory is that AIDS originated in Africa only decades in the past. Now every continent is affected. No where is the death rate from AIDS more accelerated than on the continent of Africa. It has been determined AIDS is transmitted from one person to another through blood contact and sexual practices. Furthermore, when we look at cultural practices in parts of Africa we find much prostitution and promiscuity. These behaviors are largely attributed to poverty and low levels of economic development. As a result, higher developed countries of the world have held summit meetings to initiate funding those parts of the world to increase education and economic development. Having said that, let me get right to the point of this chapter, which is dolphin mortality due to cultural practices.

Japan has been Japan for a long time; that doesn't mean that in that time they have figured out the right way to do things. At Iki, the islanders hold an annual "round-up." Here in the states when we hear the words "round-up" we usually think of the gathering of ranch animals where we ready the livestock for the up coming rodeo, but when they say "round-up" in Iki they mean round-up the dolphins and slaughter them.

The Japanese fishermen have a strange idea that when they have a bad fishing day or month or year it is because the dolphins have eaten all the fish in the ocean. Therefore, they think they need to kill all the dolphins so they can have all the fish and the dolphins get none. This is an absurd, preposterous, and sad idea! It would make more sense to say, perhaps, Japanese are stupid, inadequate fishermen.

Try to picture it in your mind and I'll try to be as graphic as I can because the only way they are going to change this absurd behavior is by pressure from outside their country. They believe in what they do, no matter how much or who thinks they are wrong:

All the little Japanese fishing men and their families gather for a ceremony to call out to the ancient Japanese gods to bless them in their task. They jump in their little fishing boats and head out to sea, taking with them any and everything they can find that makes loud noise. Wave after wave of these little Japanese in their boats go out seeking pods of dolphins, and they don't care what kind just dolphins. They herd them together and chase them in to a cove at shore where women and children await their arrival. Then man, woman, and child pick up their clubs, spears, swords, butcher knives, or pocket knives and begin the slaughter. No dolphin is to remain alive! (I'm sorry. I can't write

> this without tears. . . the image is evil and very disturbing!) A ten year old boy runs his blade into the side of a mother dolphin and spills her guts into the sea while the new-born baby awaits its fate at her side. The dolphins are quivering and splashing about as they are ***murdered*** one by one. The water in the cove turns from beautiful blue to a vulgar blood-red.

Do you get the picture?

There is no doubt that culture has made the Japanese continue this murderous slaughter generation after generation. It does not matter how long the Japanese people have been on this Earth, the dolphins have been here a long time before them. Furthermore, it is a very **stupid** assumption that dolphins eat all the fish in the ocean. Moreover, the dolphins have more of a right to the fish than humans do if, that is, it came down to one or the other having the right to the fish! God has placed far more fish in the oceans than any of the Earth's inhabitants could consume.

There is a different, yet just as disturbing, story concerning the Japanese people and dolphins—**they eat dolphins**! That's right, dolphin meat is sold in the street markets and in restaurants as a delicacy. I once saw a documentary where the reporter asked a Japanese Official if he thought they ought to be eating dolphins, and his reply was something to the effect of: *You Americans eat cows and chickens, what's wrong with us eating dolphins? You need to just mind your own business and leave us alone.*

I'll tell you what's wrong with that! Dolphins are sentient creatures, placed on the Earth by God to play a vital role in all of the Earth's oceans and bays. Dolphins have never in history been the enemy of humans, but there are numerous stories of dolphins helping humans in times of need. They are the "minds" of the sea. In some cases, they are endangered of becoming extinct. They are wild animals around the world, not bred for consumption by humans.

There is only one God! He created everything that was created. He knows how our flesh body works, so He gave us instruction on which animals to eat and which ones not to eat in the eleventh chapter of Leviticus. According to God, it is wrong to eat dolphins. Do you believe God or man?

Cows and chickens are not sentient creatures. They are domesticated and bred for human consumption. They are not endangered species anywhere in the world and humans around the whole world eat them, even Japanese. God gave us cows and chickens to consume. Read Leviticus 11.

If I were stranded on a desert Island, consisting of nothing but sand with no vegetation or wildlife, I would soon approach starvation. I would continue to ask God for food. He would probably send me a dolphin, but here's the difference . . . the dolphin would not be for me to eat, the dolphin would bring me a fish to eat! If I eat the fish and the dolphin, afterwards I would still die from starvation because God would not send me another dolphin since I abused the first one!

Culture is the difference!

14 FOOD FOR THOUGHT

Yes, dolphins are some very smart critters. You may not see them all the time, but they're out there swimming around doing their dolphin thing in every ocean of the world. They've always been dolphins, since the time God created them and placed them in the oceans. They have since procreated kind after kind. Generation after generation they are born into families with brothers, sisters, cousins, aunts, uncles, even grandparents. They live their entire lives in the water for ten, twenty, forty years, or more. Every few minutes they have to break the surface of the water, stick the top of their head into our part of the world, and breathe some of the same air you and I breathe.

God has blessed dolphins with divine peace; you see, they are not burdened with the task of worrying what happens to humans, as we are with them. For it is the human that is called to be the "steward" of God's creation. Even then, the "stewardship" call does not go out to humanity as a whole, but rather to the human as an individual.

God changes humanity through one person at a time. Perhaps, you could be one of the few God uses.

I never really considered the life of a dolphin until I reached the age of thirty-five. Back then, my spirit was on fire, and the more I learned about them the more it burned. Magazine after magazine, research paper after research paper, documentary after documentary, I became more and more sensitive *to the plight of dolphins, when humans get in the way*. What I felt was a burning desire to make humans leave dolphins alone to live their lives the way God intended them to. I wrote a hundred letters and made a hundred phone calls, some of which had the desired affect I was after, most just went by the wayside.

Nonetheless, I continued to hoe my row, so to speak. I can remember going into the library at the University of Texas at two o'clock in the afternoon and walking out the door at ten o'clock at night. I'll admit it—I was obsessed with the *plight of dolphins*! The long hours and the hard work paid off, though, because I learned a whole lot about these wonderful animals and met a lot of interesting people, some of which were important and influential to my cause.

I can honestly say that the most precious times I spent were the many hours sitting on the rip-rap of the shores of the ship channel in Port Aransas, Texas quietly watching wild dolphins doing their thing. That is the best way to learn the truth about dolphin behavior. You don't have to go to them. If you just sit there in one spot, eventually, they will come by you. In that way, you can see them playing and breaching, feeding and sleeping, fighting and mating. You can learn to identify individuals and groups. You can watch a group together, split apart, then come together again.

You should realize when you see a scientist or a chartered party boat on television following or swimming in the water near dolphins, those times are rare and undoubtedly the result of someone's long hours of study and research. There is no law against watching dolphins, but remember, within the federal waters of the United States it is against the law to feed, touch, harass, or harm a dolphin or any other marine mammal!

If you are a novice, concerning dolphins, educated or not, and want to learn more about them try the methods discussed in chapter 8. Be patient and be kind. Most of the time, a person learned in a subject is willing to share information as long as their work is not threatened or their trust betrayed.

After all of my long hours and hard work, I finally learned the very most important lesson: It is God alone, who has the ultimate say-so when it comes to *the plight of dolphins*, and *when humans get in the way* of His plan . . . He can move them out of the way!

APPENDIX A
GOVERNMENT OFFICES

President [full name]
The White House
1600 Pennsylvania Ave., NW
Washington, DC 20500

The Honorable [full name]
United States Senate
Washington, DC 20510

The Honorable [full name]
U.S. House of Representatives
Washington, DC 20515

The Honorable [full name]
Secretary of Commerce
Department of Commerce
Washington, DC 20230

[full name]
[title]
National Marine Fisheries Ser.
1335 East-West Highway
Silver Spring, MD 20910

U.S. Marine Mammal
Commission
1625 I Street, NW
Washington, DC 20006

Naval Command, Control
And Ocean Surveillance
Center
RDT&E Division
San Diego, CA 92152-5001

South Atlantic Fishery
Management Council
One Southpark Circle, #306
Charleston, S.C. 29407-4699

Animal Care Staff/Vet. Ser.
Animal & Plant Health
Inspection Services
U.S. Dept. of Agriculture
6505 Belcrest Rd., Fed.Bldg.
Hyattsville, MD 20782

(Observer Program)
National Marine Fisheries Ser.
NOAA
7600 Sand Point Way, NE
BIN C15700 - Bldg. 4
Seattle, WA 98115-0070

National Technical Info. Ser.
U.S. Dept. of Commerce
Springfield, VA 22161 Texas

Marine Mammal Stranding
Network
4700 Ave. U, Bldg. 303
Galveston, TX 77551

Marine Mammal Stranding
Ctr.
P.O. Box 773
Brigantine, NJ 08203

APPENDIX B
CETACEAN ADVOCATES

Project Interlock
P.O. Box 20
Whangarei, New Zealand
www.wadedoak.com

Earth Island Institute
2150 Allston Way
Suite 460
Berkley, CA 94704-1375
www.earthisland.org

Center For Marine Conservation
1725 DeSales Street, NW
Washington, DC 20036

American Cetacean Society
P.O. Box 1391
San Pedro, CA 90731

GreenPeace
702 H Street, NW
Suite 300
Washington, DC 20001
www.greenpeace.org

Dolphin Project
Earth Island Institute
2150 Allston Way Suite 460
Berkley, CA 94704-1375
www.dolphinproject.org

Marine Conservation Institute
4010 Stone Way N, Suite 210
Seattle, WA 98103
www.marine-conservation.org

BIBLIOGRAPY

Caldwell, M.C.; Caldwell, D.K. "The whistle of an Atlantic bottlenose dolphin (Tursiops truncatus)." Sayigh, Tyack, Wells, Scott. Signature whistles of free-ranging bottlenose dolphins: stability and mother-offspring comparisons. vol.26, no.4, pp.247-260; 1990.

Carson, Rachel L. Silent Spring. Houghton Mifflin; New York, N.Y. 1962.

Darwin, Charles. The Origin of Species. Macmillan; New York, N.Y. 1962.

Dawson, S.M.; Thorpe, C.M. A Quantitative Analysis of the Sounds of Hector's Dolphin. Ethology; vol.86, no.2, pp.131-145. Paul Pary Scientific Publisher, Berlin and Hamburg; 1990.

Diercks, K.J. "Biological sonar systems: a bionics survey." Norris and Mohl. CAN ODONTOCETES DEBILITATE THEIR PREY WITH SOUND? The American Naturalist; vol.122, no.1; July, 1983.

Doak, Wade. Project Interlock International Newsletter. July, 1992: Number 62, p.3.

Dobbs, Horace. FOLLOW THE WILD DOLPHINS. St. Martin's Press; New York, N.Y.: 1982.

Duxbury, Alyn C.; Duxbury, Alison B. An Introduction to the World's Oceans: Second Edition. Dubuque: WM. C. Brown, 1989.

Evans, Peter G.H. The Natural History Of Whales And Dolphins.

Facts On File, Inc.: New York, NY, 1987.

"Fish-Killing Source Told." The Austin Statesman 17 Jan. 1961: p.1-2.

Herman, L.M.; Morrel-Samuels, P. "Knowledge Acquisition and Asymmetry Between Language Comprehension and Production: Dolphins and Apes as General Models." Beckoff, M.; Jamison, D. [eds.] Interpretation and Explanation in the Study of Animal Behavior - Volume 1: Interpretation, Intentiality and Communication. Westview Press; Boulder, San Francisco, Oxford: pp 283-312, 1990.

Herzing, Denise. "Underwater and Close Up With Spotted Dolphins." WHALEWATCHER Journal Of The American Cetacean Society Fall 1990; Vol. 24, Number 1: 16.

Hubbs, C.L.; Rechnitzer, A.B. (1952). "Reports on experiments designed to determine the effects of underwater explosions on fish life." Norris and Mohl. CAN ODONTOCETES DEBILITATE THEIR PREY WITH SOUND? The American Naturalist. vol.122, no.1; July, 1983.

Kaplan, J. DAY OF THE DOLPHINS. Omni; p.44, 96; June, 1989.

Kellogg, W.N. PORPOISES AND SONAR. The University Of Chicago Press. Chicago, 1961.

Linehan, Edward J. "The Trouble With Dolphins." National Geographic VOL. 155, NO. 4; April 1979: 528.

Livermore, B. WATER WINGS. Sea Frontiers; March/April, p. 47-49; 1991.

Mathews, L.H. 1938. "THE SPERM WHALE: Physeter catadon". Norris, K.S.; Mohl, B. CAN ODONTOCETES DEBILITATE

THEIR PREY WITH SOUND? The American Naturalist; vol.122,no.1; July, 1983.

Mchugh, M.B. POPULATION NUMBERS AND FEEDING BEHAVIOR OF THE ATLANTIC BOTTLENOSE DOLPHIN (TURSIOPS TRUNCATUS) NEAR ARANSAS PASS, TEXAS. Master of Arts Thesis. University of Texas at Austin; May, 1989.

Murchison, A.E. "Maximum detection range and range resolution in echolocating bottlenose dolphins (Tursiops truncatus). Norris and Mohl. CAN ODONTOCETES DEBILITATE THEIR PREY WITH SOUND? The American Naturalist; vol. 122, no.1; July, 1983.

Norris, K.S. (1964) "Marine Bio-acoustics." Location Of An Acoustic Window In Dolphins. Popov, V.V.; Supin, A.y. Ethology, Ecology, Evolution. vol. 46 no. 1, 1990; 53-56.

Norris, K.S.; Harvey, G.W. 1972. "A THEORY OF THE FUNCTION OF THE SPERMACETI ORGAN OF THE SPERM WHALE (Physeter catadon)." Norris, K.S.; Mohl, B. CAN ODONTOCETES DEBILITATE THEIR PREY WITH SOUND? The American Naturalist; vol.122, no.1; July, 1983.

Norris, K.S.; Dohl, T.P. The behavior of the Hawaiian spinner dolphin (Stenella longirostris). United States National Marine Fisheries Service: Fish Bulletin; 77: pp. 821-849; 1980.

Norris, K.S.; Mohl, B. CAN ODONTOCETES DEBILITATE THEIR PREY WITH SOUND? The American Naturalist; vol.122, no.1; July, 1983.

Norris, Ken. In - NATURE. The Nature Series; The Nature Conservancy: 1992. Copyright 1992, THIRTEEN/WNET-TV.

O'Hara, Kathryn J.; Iudicello, Suzanne; Bierce, Rose. A Citizen's Guide To Plastics In The Ocean: More Than A Litter Problem. Washington, DC: Center For Marine Conservation, 1988.

Popov, V.V.; Supin, A. Y. Location of an acoustic window in dolphins. Experentia; vol.46, no.1, pp.53-56; 1990.

Postlethwait, John H., Hopson, Janet L. The Nature of Life. McGraw-Hill, New York; 1989.

Sayigh, L.S.; Tyack, P.L.; Wells, R.S.; Scott, M.D. Signature whistles of free-ranging bottlenose dolphins (Tursiops truncatus): stability and mother-offspring comparisons. Behavioral Ecology and Sociobiology; vol.26, no.4, pp.247-260; 1990.

Turbush, Ann D. Chief, Permits Division-National Marine Fisheries Service. Personal correspondence; March, 1994.

Index

E

F

G

H

I

J

K

L

M

N

O

P

Q

R

www.ingramcontent.com/pod-product-compliance
Lightning Source LLC
Chambersburg PA
CBHW051648170526
45167CB00001B/378

* 9 7 8 1 4 9 9 2 4 1 0 8 2 *